T0135610

Dynamics and Synchronization Phenomena of Semiconductor Lasers with Delayed Optical Feedback:

Utilizing Nonlinear Dynamics for Novel Applications

Vom Fachbereich Physik
der Technischen Universität Darmstadt

zur Erlangung des Grades
eines Doktors der Naturwissenschaften
(Dr. rer. nat.)

genehmigte Dissertation von
Dipl.-Phys. Michael Peil
aus Hanau am Main

Referent: Prof. Dr. W. Elsäßer
Korreferent: Prof. Dr. F. Kaiser

Tag der Einreichung: 13.01.2006
Tag der Prüfung: 22.02.2006

Darmstadt 2006
D17

Bibliografische Information Der Deutschen Bibliothek

Die Deutsche Bibliothek verzeichnet diese Publikation in der Deutschen
Nationalbibliografie; detaillierte bibliografische Daten sind im Internet über
http://dnb.ddb.de abrufbar.

ISBN 3-8325-1195-4

Logos Verlag Berlin
Comeniushof, Gubener Str. 47,
10243 Berlin
Tel.: +49 030 42 85 10 90
Fax: +49 030 42 85 10 92
INTERNET: http://www.logos-verlag.de

Contents

Chapter 1

Introduction and Overview

More than a century ago, when Poincaré worked on the three-body problem, he reckoned the existence of extremely complex trajectories which he denoted as *"trajectories that are so complex that I did not intend to draw their pictures..."* [1]. With this finding, he gave a first indication for complex dynamical behavior. In that sense the 1890s might be referred to as the years of discovery of *Chaos* and *Complexity*. However, it took more than 60 years, and the development of computers, until Lorenz, and Henon and Heiles independently recognized the fundamental unpredictability of such complex motions produced by deterministic models consisting of coupled, nonlinear, ordinary differential equations. Lorenz's work on non-periodic flows that occurred in his model of the earth's atmosphere [2], and Henon's and Heiles' research on motion of stars in the galaxy [3], showed that solutions of deterministic models can exhibit extreme sensitivity against small changes of the initial conditions resulting in exponentially diverging nearby trajectories. This pioneering work on *Deterministic Chaos* marked a breakthrough in the understanding of nonlinear systems and eventually initiated this rapidly growing scientific field of *Nonlinear Dynamics* (NLD) and *Chaos*. In the following twenty years, much attention has been paid to the emergence of chaotic behavior, and four universal routes to chaos have been identified, namely, the quasi-periodic route [4], the period-doubling cascade [5], intermittency [6], and crisis [7]. Contemporarily, experiments on various systems in diverse areas of science, such as in biology, chemistry and physics, have shown the occurrence of chaotic dynamics, e.g., in blood cell production, neurons, population dynamics, chemical reactions, mechanical pendula, fluid flows, and many different laser systems [8,9]. This, and other, experimental work has not only verified the theoretically predicted routes to chaos but also disclosed general principles underlying nonlinear dynamics. This discovery is considered as one of the most important scientific discoveries of the last forty years. From that on, many dynamical phenomena have been revealed and quantitative methodologies for characterization of nonlinear dynamics have been developed, e.g., strange attractors, embedding theory, dimension measures, symbolic dynamics, Lyapunov exponents, entropy, and many other information theory-based measures [8–10].

1

From the early years on, the development of NLD has been substantially supported by experiments in *Optics*. The close connection between both scientific fields first became clear in 1975, when Haken discovered isomorphism between the Lorenz equations and the Maxwell-Bloch equations describing lasers [11]. This finding demonstrated that laser systems possess an inherent instabilities which were soon experimentally verified for gas lasers [12–15] and solid state lasers as well [15–18]. These results initiated many experiments utilizing laser systems as prototypes for studying NLD phenomena. In particular, the observation of instabilities and chaos in widely used semiconductor lasers was of importance [19], because it has accelerated research on NLD and chaos in optics and, thereby, closely linked the development of both scientific fields. Today, the field of nonlinear optics, including nonlinear dynamics in laser systems, is an integral part of modern optics.

The history of *Semiconductor Lasers* (SLs) is as comparably young as that of NLD. In similarity to NLD, *Semiconductor Laser Physics* has undergone a tremendous development since realization of the first SL devices in 1962 [20–22]. While the first generations of SLs were quite impractical devices, which only allowed for pulsed operation under liquid nitrogen temperature conditions. Improved structures, continuous development of device growth and processing techniques facilitated significant improvements of SL performance. Today, the emission properties of SLs can be tailored by influencing the material composition, the structure and the geometry of the device. Subsequently, the optimized SLs can be grown epitaxially on wavers allowing for cost efficient industrial mass production. Modern devices operate continuous wave at room temperature for many thousands of hours without significant degradation. In particular, state-of-the-art SLs are highly efficient light sources with wall-plug efficiencies which can exceed 60 %. Additionally, they allow for high frequency modulation of up to several tens of GHz. Their extraordinary small physical dimension, filling a volume of less than 0.1 mm^3, offers great potential for integration into electronic devices, or even for production of monolithic micro electro-optical compound devices. These particular advantages of SLs, if compared to conventional light sources, have facilitated that SLs have become established electro-optical key elements in a great variety of applications in consumer electronics industries and research. In consumer electronics SLs are implemented in CD and DVD data storage and read-out systems, in bar code scanners, and in data and telecommunication technology; in industries SLs are utilized in sensor technology, ranging applications, and for pumping of solid state lasers, while modern high-power SLs even allow for material cutting and welding. In research, SLs are mainly applied in laser cooling experiments in quantum optics, or as coherent tunable light sources for spectroscopy. Besides these applications, the excellent properties of SLs have attracted particular interest in the field of NLD. The combination of their inherently strong non-linearity, their pronounced sensitivity to external perturbations, and their potential for high frequency intensity dynamics predestine SLs as experimental model-systems for studying fundamental NLD phenomena. In fact, SL systems have proven their high potential for experimental studies of fundamental questions of NLD, and they have contributed significantly to today's understanding of the emergence of the classical routes to chaos, i.e., via intermittency [23], bifurcation cascades [24], period-doubling [25, 26], and quasi-periodicity [25, 27].

Meanwhile, many of the fundamental properties of NLD in low dimensional systems are well studied and understanding of the phenomena has reached a state of maturity, while investigated NLD in high-dimensional and spatially extended systems revealed new phenomena, such as self-organization [10, 28], dynamical pattern generation [29], or spatiotemporal chaos [30]. In fact, actual experiments on high-dimensional systems continuously bear new, exciting phenomena which are not always captured by the existing methodology of NLD. Furthermore, the occurrence of high-dimensional nonlinear dynamics is a wide spread phenomena, since it can easily emerge from coupling of many low-dimensional nonlinear oscillators forming a network, or it can be induced by time-delayed feedback in single nonlinear systems, while combinations of both effects are also possible. Therefore, research on high-dimensional systems is of general importance in diverse scientific fields.

The human brain represents the most prominent example of the multitude of high-dimensional nonlinear systems in nature. It belongs to the class of complex networks and consists of numerous coupled nonlinear oscillators, called neurons which can even exhibit significant delays in their coupling. In the human brain, and in many other high-dimensional and also low-dimensional, nonlinear systems, spatiotemporal synchronization effects are relevant phenomena influencing functionality of the system. Today, synchronization has been identified as being an essential phenomenon in nature; even more, it is seen as one of the most mysterious and pervasive drives in nature [31, 32]. For that reason, NLD is one of the most active, and attractive, interdisciplinary scientific fields. Nowadays, the main interest in NLD is dedicated to the investigation of the generic properties of high-dimensional nonlinear systems, often referred to as *Complex Systems*, while special attention is being paid to the investigation of synchronization effects. Consequently, a holistic understanding of the origins, emergences, and effects of the various dynamical phenomena in complex real-world systems is desired and mainly motivated by the following reasons:

- From the fundamental point of view, the universality of high-dimensional nonlinear dynamical phenomena requires profound, interdisciplinary investigation. Since analysis of high-dimensional nonlinear systems is typically very challenging, available methodology of NLD can serve as basis for investigation and description of complex structures in space and time. Advancing research continuously reveals limits of existing methodology which boosts development of NLD. Adapted methodology, in turn, will give further insight into generic mechanisms of high-dimensional nonlinear dynamics in diverse complex real-world systems.

- From the applications point of view, knowledge of generic mechanisms influencing complex behavior will allow for precise manipulation and control over systems exhibiting chaotic dynamics. As a consequence, high-dimensional chaos can be harnessed to achieve some practical goals, such as prediction, control, and chaos communication.[1]

Unfortunately, complete experimental analysis of the dynamics of most complex real-world systems is presently beyond scientific possibilities. This is due to several reasons. Often, relevant system parameters are not accessible, or sufficiently accurate control over the dynamics is lacking. Furthermore, appropriate detection of the dynamics cannot be achieved; and in many real-world systems, the relevant time scales of the dynamics are comparably slow so that the dynamics is influenced by changing environmental conditions, e.g., in biological and chemical systems. Analytical investigation of appropriate models, which has proven to be a powerful technique in low-dimensional systems, could provide an alternative. But in contrast to low-dimensional systems, analytical treatment of models capturing the dynamics of only moderately complex high-dimensional nonlinear systems is very demanding and often not feasible. Fortunately, within the last decade, new generations of computers with sufficiently improved performance have been developed. For the first time, these computers allow for numerical access to moderately high-dimensional nonlinear systems. However, many complex systems exhibit dynamics which still is beyond the potential of modern computers, impeding detailed analysis of various essential dynamical phenomena, which remains a major challenge. Therefore, experimental model-systems provide a promising alternative to obtain insight into universal dynamical phenomena of complex dynamical systems. Such appropriate model-systems should provide well-controllable systems parameters; well-accessible dynamics, generated under constant environmental conditions; experimental versatility, and potential for varying the degree of complexity. Furthermore, possibility for complementary modeling is desired.

Semiconductor lasers which are subject to delayed optical feedback are well-suited model-systems for studying high-dimensional NLD phenomena. This becomes clear from the following reasons. On the one hand, SLs are experimentally well under control. They have been intensively studied in the last decades, and, meanwhile, their dynamical properties are well understood. Based on the gained knowledge, efficient

---

[1] An introduction to utilization of chaos is given in the book "Coping with Chaos" by E. Ott [33].

models have been developed which accurately mimic their dynamical behavior. Major distinction of SLs originates form the characteristic combination of the inherently strong nonlinearity and the high sensitivity of the emission properties to external perturbations. Both are ideal premises for generation of pronounced nonlinear intensity dynamics, which for SLs can be fully captured by application of state-of-the-art measurement technology. Particularly intriguing are SLs which are perturbed by applying delayed optical feedback. Such ECSL systems exhibit various high-dimensional NLD phenomena, while the dynamics is well-accessible allowing for thorough studies thereof. On the other hand, coupled SL feedback systems are well-suited for investigation of diverse fundamental synchronization phenomena, since complexity and topology of the "model-networks" can be adjusted by variation of the number of coupled oscillators and their coupling-configuration.

In this thesis, we explore the strong linkage between NLD and SL physics. Our motivation for this is twofold. On the one hand, we are interested in experimentally studying fundamental NLD phenomena by using SL systems. Here, we focus on investigation of high-dimensional chaotic dynamics and synchronization phenomena. We apply methods of NLD to gain insight into fundamental mechanisms underlying these phenomena. On the other hand, such insight could, in turn, be utilized for controlled manipulation of these phenomena and the connected emission dynamics. Consequently, we follow this strategy and investigate the possibility of tailoring the emission properties of SLs and of optimizing the synchronization properties of unidirectionally coupled SL systems. We verify that such tailoring of nonlinear dynamical properties is indeed possible. Furthermore, our results demonstrate that NLD can be utilized to exploit the unique attributes of chaos and synchronization for practical use.

This present work is organized as follows. In Chapter 2, we discuss the conceptual meaning of semiconductor lasers and NLD. We elucidate the close linkage between both scientific fields and highlight numerous advantages of SLs, motivating their usage for experimental studies of fundamental NLD phenomena. To provide the necessary theoretical background, we outline the mathematical description of lasers and describe how, under consideration of the particular features of SLs, an applicable rate-equation model can be derived. In this context, we discuss important dynamical properties of SL systems, such as their inherently strong nonlinearity and their distinct susceptibility to external perturbations, with respect to investigation of intriguing high-dimensional NLD phenomena.

The experimental investigation of general properties of high-dimensional NLD phenomena represents one integral part of this work. Therefore, in Chapter 3, we extensively study the emission properties of SLs with delayed optical feedback, with focus on high-dimensional, broadband, chaotic intensity and spectral dynamics. We apply

state-of-the-art measurement technology and methodology of NLD to gain insight into
the fundamental dynamical phenomena within this regime, which have so far been
beyond experimental accessibility. Reason for this can be found in the wide range of
spectral and temporal scales which are involved in the dynamics. We demonstrate
that the dynamics sensitively depends on relevant systems parameters allowing for the
controlled generation and manipulation of high-dimensional chaotic dynamics. Furthermore, we reveal a close interrelation between the intensity and the spectral dynamics
which can be harnessed for tailoring the coherence properties of SLs via influencing the
dynamical properties of the system. In this context, we demonstrate that SLs systems
for which the delay time is shorter than the period of the relaxation oscillations offer
excellent performance, since they exhibit versatile dynamical phenomena comprising
well-pronounced broadband intensity chaos. Hence, these short ECSL systems represent attractive experimental systems for fundamental NLD, while the high versatility
of the dynamics explicitly offers potential for novel applications.

In Chapter 4, we apply the obtained knowledge of the nonlinear dynamical properties
of SLs with delayed optical feedback for tailoring the coherence length of SLs. Our
goal is to drastically reduce the coherence qualities of SLs, since coherence effects often
restrict the resolution of metrology applications of SLs. We have chosen a particularly
long edge-emitting SL and designed the external cavity of the SL system to influence
its dynamical properties. We adjust for comparable lengths of the external cavity
and the SL cavity and realize simple resonance conditions between both cavities to
enhance the coupling between the LMs. For these conditions, we vary the relevant
system parameters and find dynamics which reveal distinct differences if compared to
the known dynamics of conventional non-resonant short cavity systems. In particular,
we demonstrate that it is possible to achieve spectral broadband emission for which
the number of lasing LMs can easily exceed 130 covering a spectral bandwidth of approximately 7 nm. We analyze the spectral properties of this pronounced multimode
dynamics and find considerably reduced coherence properties with minimal coherence
lengths on the sub-millimeter scale. Such successful reduction of the coherence length
offers high potential for ECSL light sources for implementation in modern measurement applications. Hence, our results verify that NLD can be successfully applied for
tailoring the emission properties of SLs, while the tailored ECSL system proves to be
a versatile and attractive experimental model-system for studying general properties
of pronounced multimode dynamics.

The third integral part of this thesis is dedicated to analysis of synchronization phenomena of coupled delay systems. In this regard, SLs with delayed optical feedback also
prove to be beneficial experimental model-systems for studying fundamental synchronization phenomena of complex dynamical systems. In Chapter 5, we concentrate on

investigation of the synchronization properties of unidirectionally coupled chaotically emitting SL systems. In addition to our motivation from the fundamental NLD point of view, we are especially interested in whether it is possible to harness synchronization properties of SL systems for the realization of a functional encrypted communication system. As a starting point, we introduce the basic principle of *Chaos Communications* and discuss the requirements on the carrier signal and the synchronization properties of a functional chaos communication system. We show that the operation conditions of ECSLs can be optimized to generate well-suited broadband chaotic carrier signals allowing for good concealment of a message. We verify that excellent "chaos" synchronization can be achieved for two different configurations of receiver systems, the *Open Loop* and the *Closed Loop* configuration. We investigate the synchronization properties of both configurations and evaluate the properties in terms of security aspects and practicability for chaos communications. In this context, we consider different encryption and decryption techniques, which are based on particular synchronization properties and which determine the reliability and the security of the communication system. Based on these considerations, we present a functional chaos communication system for which we demonstrate successful transmission of a test signal. Finally, we summarize the results and compare the characteristics of different configurations providing an outlook for possible improvements of future constructive applications of chaos synchronization.

We conclude this thesis with a summary of the main results emphasizing the potential of harnessing NLD for novel functionalities.

Chapter 2

Semiconductor Lasers and Nonlinear Dynamics

In the past three decades, *Semiconductor Lasers* (SL) have become vital electro-optical devices for various applications in industries and everyday life. Today's SLs-industry has evolved into the economically most significant laser branch. This fact is based on the beneficial properties of SLs comprising small physical dimension, cost-effective production, and highly efficient conversion of current to coherent light. In particular, SLs exhibit good emission properties, being typical for a laser, that can be well controlled and optimized depending on the demands of the particular technical applications in, e.g., CD/DVD data storage and read-out systems, laser printers, material processing, pumping of solid-state lasers, and, last but not least, in modern telecommunications. However, SLs also react sensitively even to small external perturbations, such as optical feedback, optical injection and current noise, resulting in modifications of the emission properties comprising, e.g., linewidth, output power, and wavelength stability. Therefore, optimization of the emission properties of SLs, which is desired for continuously developing technology and applications, requires consideration of the dynamical properties of the laser. To achieve this goal, profound understanding of the emission properties of SLs is required. In this context, a detailed physical description of the relevant mechanisms underlying the emission properties is desired. However, in contrast to the practical operation of SLs, the physical description of the lasing process is not a trivial task because it involves complex, nonlinear interaction processes between optical field and matter in the SL cavity [34].

This nonlinear interaction of light and matter is a generic property in all laser systems, while the time scales of the underlying mechanisms can vary depending on the particular type and structure of laser. The time scales, in turn, define the dynamical properties of the laser, in large. This generic property of lasers indicates the conceptual connection between laser physics and *Nonlinear Dynamics* (NLD), which has been verified by Haken in 1975 who demonstrated that the Maxwell-Bloch equations describing laser systems are mathematically isomorphic to the Lorenz equations [11]. For that reason, the development of lasers, and, in particular, that of semiconductor lasers, is closely linked to NLD. Reciprocally, SLs have proved their value for studying general NLD phenomena, e.g., intermittency [23], bifurcation cascades [24], period-doubling [25, 26], quasi-periodicity [25, 27], and also high-dimensional dynamical phenomena have been studied, recently [35–37].

To date, investigation and characterization of high-dimensional chaotic dynamics is still a challenging task. Nevertheless, understanding of generic properties of high-dimensional dynamical systems is desired, since it allows for control of the dynamics which can be utilized in two different ways. On the one hand, such understanding allows for controlled generation of high-dimensional chaotic dynamics with tailored properties, and on the other hand for suppression of such dynamics, while both possibilities open up potential for new applications.

In this context, laser systems and, in particular SL systems, are ideally suited for experimentally studying high-dimensional nonlinear dynamics, since they offer excellent properties, if compared to other experimental systems, e.g., in biology, in chemistry, or in fluid dynamics. SL model-systems offer four main advantages. Firstly, most lasers are well studied devices with well-known, and well-accessible parameters. Therefore, they are experimentally well under control which allows for operation under defined experimental conditions. Secondly, laser systems offer a hierarchy of relevant time scales that, depending on the type of laser, can span many orders of magnitudes, typically, in a range between femtoseconds and milliseconds time scale. This property is beneficial for two reasons. On the one hand, even the slow time scales are sufficiently fast to allow for acquisition of long time series without significant variations of the systems parameters. This offers potential for application of methods for quantitative time series analysis, i.e., determination of information theory-based measures. On the other hand, the great hierarchy of time scales in laser systems provides a fruitful basis for the emergence of versatile dynamical phenomena. Thirdly, lasers systems are constrained systems for which the number of dynamical degrees of freedom can be well controlled. Fourthly, laser systems belong to the class of dissipative systems, because of light emission and thermal losses. This property is of great advantage for experiments as well as for modeling, because consideration of transient dynamics is not essential for characterization of general "long-term" dynamical properties. This results from the fact that dissipation "blanks out" memory about initial conditions when the dynamics converges to asymptotic solutions, called *Attractors*, which are relevant for characterization of the dynamical properties.

These advantages of laser systems point out their potential for investigation of fundamental aspects of NLD, since laser systems allow for studies under experimentally well-controllable and well-defined conditions. Therefore, laser systems have become an essential model-system of experimental NLD. Based on this conclusion, in this chapter, we motivate for the utilization of SLs for experimental studies of high-dimensional chaotic dynamical phenomena, since this particular type of laser offers distinct beneficial properties. For discussion of the particular properties of SLs, we outline the mathematical description of laser systems, in general, and introduce approximations leading to a simplified rate-equation model. We discuss the model in terms of the relevant time scales of the underlying physical processes and show how the model can be adapted to capture the emission properties of SLs by considering the relevant physics and the corresponding time scales. This procedure leads to the well-known semiconductor laser rate-equations, from which we will later deduce numerical models for complementary analysis supporting our experimental studies. Accordingly, we discuss the SL rate-equations and elucidate particularities of SLs, which are relevant for their NLD properties. In this context, we briefly introduce the different SLs which are utilized for the experiments presented in this thesis and classify their dynamics characteristics and peculiarities.

2.1 Modeling the Emission Properties of Lasers

In this chapter, we outline the derivation of a mathematical model describing the interaction between light and gain medium in a laser cavity. The model provides insight into the emission properties of lasers from a theoretical point of view. Additionally, it is the starting point for the subsequent derivation of the SL rate-equations on which modeling of our experiments is based on.

For derivation of a model which describes the emission properties of lasers it is essential to consider the interactions between an "intense" electromagnetic field and matter, representing the gain medium. In a first step, a description of the physical process of the light-matter interaction in the gain medium is desired. For a two-level system, such a description can be derived quantum mechanically from the Schrödinger equation, in terms of density matrix elements [38]. The result relates the electrical field with the microscopic material polarization and the population inversion and allows for determination of the macroscopic polarization. In a second step, the relation between the electrical field $\mathbf{E}(\mathbf{x}, t)$ and the macroscopic polarization of the gain material $\mathbf{P}(\mathbf{x}, t)$ which leads to the lasing process needs to be described. This relation is captured by Maxwell's equations [39] which need to satisfy the boundary conditions defined by the laser cavity:

$$\nabla^2 \mathbf{E}(\mathbf{x}, t) - \frac{\sigma}{\epsilon_0 c^2} \frac{\partial}{\partial t} \mathbf{E}(\mathbf{x}, t) - \frac{1}{c^2} \frac{\partial^2}{\partial t^2} \mathbf{E}(\mathbf{x}, t) = \frac{1}{\epsilon_0 c^2} \frac{\partial^2}{\partial t^2} \mathbf{P}(\mathbf{x}, t) \qquad (2.1)$$

In this model, the classical Maxwell equations describe the optical field, while the laser medium is modeled within the frame of Quantum Mechanics. Hence, this model represents a semiclassical approach to the lasing phenomenon. In practice, a solution to Equation 2.1 can be iteratively obtained in the following procedure. First, a proposed solution for the electrical field $\mathbf{E}(\mathbf{x}, t)$ is inserted into the quantum mechanically derived equations describing the light-matter interaction. The equations reveal that the electrical field induces microscopic dipole moments in the gain medium from which the population inversion can be calculated. Additionally, summation over the microscopic dipole moments allows for estimation of the macroscopic polarization $\mathbf{P}(\mathbf{x}, t)$, which is required for the solution of the Maxwell equations. Hence, this estimate can be inserted into Equation 2.1 which, in turn, allows for calculation of a new solution for the electrical field $\mathbf{E}'(\mathbf{x}, t)$. For determination of the accurate solution, this procedure needs to be repeated until self-consistency between $\mathbf{E}'(\mathbf{x}, t)$ and $\mathbf{E}(\mathbf{x}, t)$ is achieved. However, this procedure can be quite intricate with respect to practical aspects. Therefore, a simplified and practicable mathematical description of lasers is desired.

2.1.1　The Maxwell-Bloch Equations

In this subsection, we address the question whether the presented model of lasers, which is based on Maxwell's equations and self-consistent solutions, can be simplified to a practical mathematical model. In order to achieve this goal, several physical assumptions and mathematical approximations can be made for simplification [40]. In the following, we outline the particular simplifications that can be made for reduction of the model leading to the well-known Maxwell-Bloch laser equations.

In a first step, we assume that the electrical field can be described as harmonic plane wave propagating in a unidirectional ring resonator. Then, we account for the fact that the optical field amplitude usually varies on time scales significantly slower than the optical phase. This property justifies application of the Slowly Varying Envelope Approximation (SVEA) [41] and allows for transformation of Equation 2.1 into a first order differential rate-equation. Furthermore, we assume that the gain medium can be described as a homogeneously broadened two-level system, and that the mirror losses can be treated as being continuously distributed over the laser resonator. Additionally, for most practical purposes it is sufficient to restrict the model to description of laser emission in one "single" longitudinal mode. Finally, very fast oscillating expressions which cancel out during integration over one optical period can be neglected, an approximation which is known as Rotating Wave Approximation (RWA) [15].

Following the above-listed assumptions and approximation, we are able to derive a model that describes the emission properties of homogeneously broadened, singlemode, two-level unidirectional ring lasers in the plane wave approximation. We obtain three differential equations for the dynamical variables, which are the electrical field $E(t)$, the macroscopic polarization $P(t)$, and population inversion $N(t)$; all of them are mutually coupled via the optical gain g. These equations are referred to as the semi-classical Maxwell-Bloch equations [15]:

$$\dot{E}(t) = -(i\omega_E + \gamma_E)E(t) - igP(t) \tag{2.2}$$

$$\dot{P}(t) = -(i\omega_P + \gamma_P)P(t) + gE(t)N(t) \tag{2.3}$$

$$\dot{N}(t) = J - \gamma_N N(t) - ig[E(t)^\star P(t) - E(t)P(t)^\star] \tag{2.4}$$

The resulting Equations 2.2-2.4 comprise relevant laser device characteristic frequencies: the resonance frequency of the laser resonator ω_E, and the frequency of the optical transition ω_P. Furthermore, the following laser specific decay rates, originating from the properties of the light-matter interaction in the laser cavity, are of particular importance from the dynamics point of view: the decay rate of the optical field in the laser resonator γ_E, often referred to as cavity decay rate; the decay rate of the macroscopic polarization γ_P; and the decay rate of the inversion γ_N. Finally, the bias current of the SL is considered in the pumping coefficient which represented by J.

These parameters are of great importance for the dynamical properties of the laser, since they result from the characteristics of the light-matter interaction in the laser cavity. We point out that there exist three characteristic frequencies which are associated with the light-matter interaction in the cavity, and which define the dynamical properties of the laser. These frequencies are: the relaxation oscillation frequency, originating from the interplay of the number of photons in the laser cavity and the population inversion; the longitudinal mode spacing, which corresponds to the frequency difference between the longitudinal modes which resonate in the laser cavity; and the Rabi precession frequency, which refers to coherent oscillations of the material polarization, induced by the applied electrical field. Under appropriate experimental conditions, and depending on the type of laser, dynamics instabilities can appear which are associated with these frequencies. The relevant time scales related to the three decay rates, γ_E, γ_P, and γ_N, which are given by the material properties of the laser, can be considered for further reduction of the Maxwell-Bloch equations, Eq. 2.2-2.4. These considerations allow for classification into three different laser categories, for which the appropriate model equations exhibit different NLD properties. In the following subsection, we address this classification into the three laser categories and briefly discuss the corresponding dynamical properties.

2.1.2   Laser Categories

Within validity of the underlying assumptions, the Maxwell-Bloch equations provide a practicable mathematical model for the emission properties of lasers. However, with respect to good analytical and numerical treatment, and in order to allow for detailed studies of generic NLD phenomena, reduction of the model according to the physical properties of the particular type of laser is desired. So far, derivation of the Maxwell-Bloch equations did not include consideration of specific material properties of the gain medium. Those define specific time scales of the physical mechanisms being involved in the interactions between light and matter in the laser cavity. These properties are included in the Maxwell-Bloch equations, Eq. 2.2-2.4, in terms of the three decay rates, which are the three decay rates of the electrical field γ_E, of the inversion γ_N, and of the polarization γ_P. Reduction of the model can be achieved, if one of the decay rates is sufficiently faster than the others. In this case, it is possible to adiabatically eliminate the corresponding variable from the system of equations reducing the number of equations by one. Based on this consideration, it is possible to classify lasers into three categories with different NLD properties. Depending on the category of laser, either one, two, or three equations are required for appropriate description of the dynamics [15]; or in other words, lasers which can be described by the Maxwell-Bloch equations are categorized according to their number of degrees of freedom. From the dynamics point of view, this result is very relevant, since it is well-known from the Poincaré-Bendixon-theorem [8, 9] that systems with two or fewer degrees of freedom cannot exhibit chaotic behavior. For such laser systems, the number of degrees of freedom needs to be increased to allow for chaotic dynamics. This can occur, e.g., via application of external perturbations.

For *Class A* lasers the decay rates of the macroscopic polarization γ_P and the inversion γ_N are sufficiently larger than the decay rate of the electrical field γ_E. Hence, the corresponding variables can be adiabatically eliminated and the dynamics is described by a single rate-equation for the electric field $E(t)$. Therefore, Class A lasers belong to the class of one-dimensional systems for which only time independent solutions exist [8, 9]. Dye lasers are famous examples of Class A lasers.

For *Class B* lasers only the decay rate of the macroscopic polarization γ_P is distinctly larger than the decay rates of the inversion γ_N and the electric field γ_E, respectively. Therefore, the dynamics of Class B lasers is captured by two rate-equations for $E(t)$, and $N(t)$. Accordingly, Class B lasers represent two-dimensional systems for which periodic solutions are possible [8, 9]. These periodic oscillations are called "carrier" relaxation oscillations of lasers, because the laser energy oscillates between inversion and optical field. Besides CO_2 lasers and microchip lasers, SLs often are treated as

Class B lasers. However, we note that this assumption is not generally justified, because of the assumptions underlying the Maxwell-Bloch rate-equation model, such as singlemode emission and neglect of spatial extension of the laser cavity. In Chapter 4, we will present a SL system, which exhibits properties that clearly violate several of the assumptions made for the rate-equation modeling approach. However, many of the conventional SLs can be classified as Class B and fulfill the requirements for application of such a modeling approach.

For *Class C* lasers the three decay rates are of similar order of magnitude. In this case, description of the dynamics requires the full set of the Maxwell-Bloch equations. Therefore, Class C lasers belong to the class of three-dimensional systems which can exhibit chaotic dynamics [8,9], if the so-called secondary laser threshold is exceeded. Indeed, in 1986, Weiss et al. have reported Lorenz-type chaos for a NH_3-laser [42]. Another well-known Class C laser for which chaotic dynamics can be observed is the He-Ne laser emitting at $3.39\,\mu$m.

Since we are interested in high-dimensional nonlinear emission dynamics of SLs, the Class B laser rate-equations represent the starting point for derivation of the model equations which we require for complementary analysis to our experiments. Certainly, we keep in mind the existing limits of validity of these equations. Furthermore, additional degrees of freedom are needed to be introduced into the dynamically two-dimensional SL system to allow for chaotic behavior. In SLs, high-dimensional chaotic dynamics can be easily generated by modulation of the pump current, i.e., by changing the inversion [43, 44]; by optical injection [43, 45]; or by application of time-delayed feedback to the relevant variables, the inversion [46] and the electric field [43, 47]. In particular, SLs which are subject to time-delayed optical feedback are very promising candidates for experimentally studying high-dimensional chaos. Before we study the influence of delayed optical feedback on the emission dynamics of SLs, we highlight characteristic features of the emission properties of solitary SLs. As starting point and as mathematical foundation, we derive SL rate-equations which describe fundamental features of the emission properties of solitary SLs.

2.2   Semiconductor Lasers

In this section, we discuss characteristic properties of SLs which are relevant for their dynamical behavior. Initially, we consider their characteristic time scales to derive a simple rate-equation model for unperturbed solitary SLs. On the one hand, the SL rate-equations will serve as basic equations for modeling feedback induced instabilities in SLs. On the other hand, analysis of the solitary SL rate-equation gives first insight into relevant physical mechanisms, and the corresponding parameters, which determine their NLD properties. We elucidate the most relevant parameters and we give insight into their dependence of experimental conditions. Subsequently, we introduce the different types of SLs, which are utilized in the experiments in Chapters 3-5, and compare their characteristics with respect to their dynamical properties.

2.2.1 Semiconductor Laser Rate-Equations

So far, we have seen that for certain physical conditions the Maxwell-Bloch rate-equations can be reduced to a one-, or two-dimensional equation system by adiabatic elimination of two, or one, of the three variables. This allowed for classification into three laser classes, depending on the specific time scales of the three decay rates of the lasers, i.e., the polarization decay rate γ_P, the decay rate of the inversion γ_N, and the decay rate of the electric field γ_E. In order to derive an appropriate model for the emission properties of SLs, it is necessary to consider the fact that electron-electron scattering destroys coherence of the coupling between the states of the conduction-band and states of the valence-band, quickly damping polarization oscillations in the semiconductor material [48]. This property is reflected by the large polarization decay rate exhibited by SLs which is of the order of $\gamma_P \sim 10^5\,\mathrm{ns}^{-1}$. In contrast to this, the decay rates of the inversion and of the electrical field are distinctly smaller: $\gamma_N \sim 1\,\mathrm{ns}^{-1}$, and $\gamma_E \sim 10^3\,\mathrm{ns}^{-1}$. Hence, the polarization reacts instantaneously to changes of the inversion or of the electrical field and can be adiabatically eliminated from the Maxwell-Bloch equations, Eqs. 2.2-2.4 [49]. As a result, we obtain two coupled rate-equations for the electric field and the inversion of the SL which take the following form for specific transformation of the variables:

$$\dot{E}(t) = \left(i\omega(N(t)) + \frac{G(N(t))}{2} - \frac{\Gamma}{2}\right)E(t) \qquad (2.5)$$

$$\dot{N}(t) = J - \gamma_N N(t) + G(N(t))|E(t)|^2 \qquad (2.6)$$

In these equations, the following parameters, describing device-specific properties and operation conditions, are relevant for appropriate modeling of the dynamical properties of specific SLs: the resonator losses Γ; the carrier decay rate γ_N; and the pumping coefficient, J, which is related to the experimental pump parameter $p = \frac{I_{DC}}{I_{th,sol}}$ as follows: $J = \frac{pI_{th,sol}}{e}$, where I_{DC} denotes the injection current, $I_{th,sol}$ the laser threshold current, and e the electron charge. Furthermore, description of the characteristic behavior of the optical gain, $G(N(t))$, and the laser frequency, $\omega(N(t))$, of the specific SL is required, both being nonlinear functions of the carrier density $N(t)$. However, for conventional operation conditions, linearization of $G(N(t))$ and $\omega(N(t))$ around the laser threshold (th) is justified:

$$G(N) = G(N_{th}) + \xi(N - N_{th}) = \Gamma + \xi\Delta N \tag{2.7}$$

$$\omega(N) = \omega_0 + \omega_N(N - N_{th}) = \omega_0 + \frac{\alpha}{2}\xi\Delta N \tag{2.8}$$

Here, ξ denotes the differential gain; while ω_0, stands for the laser frequency at threshold; and $\alpha = \frac{2\omega_N}{\xi}$ represents the so-called linewidth enhancement factor of the SL. Substitution of the linearized expressions for the optical gain, Eq. 2.7, and the oscillation frequency of the laser, Eq. 2.8, into Equations 2.5-2.6 leads to the well-known SL rate-equations. The SL rate-equations describe the nonlinear relation between the slowly varying amplitude of the complex electrical field, $E(t) = |E(t)|e^{i\phi(t)}$, and the carrier density, $N(t)$:

$$\dot{E}(t) = \frac{1}{2}(1 + i\alpha)\xi\Delta N(t)E(t) \tag{2.9}$$

$$\dot{N}(t) = J - \gamma_N N(t) - \left(\Gamma + \xi\Delta N(t)\right)|E(t)|^2 \tag{2.10}$$

This two-dimensional equation can be solved analytically, while the solutions can be given in a closed manner. Since the SL rate-equations represent a two-dimensional system, the equations only allow for periodic solutions, and not for chaotic behavior. However, a closer look at the SL rate-equation already reveals two relevant properties of SLs with respect to NLD. The first property is related to the potential for generation of periodic solutions in the system. This represents one fundamental physical mechanism for the occurrence of dynamics which consists of the coupling between the carrier inversion and the electrical field. These oscillations are known as relaxation oscillations. The second relevant property is related to the origin of the α-parameter which represents the nonlinearity of the system. This nonlinearity originates from the characteristics of the interactions between the electric field and the gain medium of the SL. Hence, from the analysis of the two-dimensional SL rate-equations, we have identified two relevant characteristics influencing the dynamical properties of SLs. We will see that the relaxation oscillations and the α-parameter are both important for the occurrence of chaotic dynamics in SL systems.

Dimensionality of dynamical systems can be increased by introducing additional degrees of freedom to the systems. If the dimension of the modified systems is three of higher, the systems distinguish themselves from the previous systems in that their dynamics can exhibit extreme sensitivity against small changes of the initial conditions. In other words, such modified systems allow for deterministic chaos. To get insight into the properties of chaotic dynamics of high-dimensional SL systems, we first need to consider the particularities of SLs. We address these particularities of SLs in the following subsection; while we pay special attention to the fundamentals of the relaxation oscillations and the α-parameter.

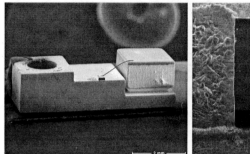

Fig. 2.1: Electron microscopy photographs of a commercially available edge emitting SL for telecommunications applications.

2.2.2 Particularities of Semiconductor Lasers

In order to give an impression of the device of interest, Figure 2.1 depicts an example of a commercially available edge emitting SL which is utilized in telecommunication applications. The left panel shows the edge emitting SL with its mount, while the right provides a magnified view on the SL. The $300\,\mu$m long front facet of the SL is located at the bottom of the photograph. Additionally, one can see the electrical contact of the SL on its top, which is realized by bonding a gold wire to the SL. It becomes clear from the figure that the device can be easily implemented into modern communication technology because of its small physical dimensions. Generally, SLs offer many advantages over other types of lasers. In addition to the practical advantages, such as high wall-plug efficiency, low power consumption, small physical dimension, cost-effective mass production, etc., SL exhibit unique qualities giving rise to interesting emission dynamics.

A first striking characteristic of SLs is their pronounced intrinsic nonlinearity which originates from a conspicuously strong coupling between the amplitude and the phase of the optical field in the SL medium. This property is expressed by the linewidth enhancement factor which is referred to as α-parameter in the SL rate-equations, Eq. 2.9- 2.10. Additionally, the relevant time scales underlying the mechanisms of the light-matter interactions in SLs are very fast, if compared to other types of lasers. Even the slowest time scale, which corresponds to the damping rate of the carrier inversion, is of the order of nanoseconds time scale which roughly defines the time scale on which the dynamics evolves. Furthermore, SLs react very sensitively to optical feedback. This is mainly due to the high optical gain of the SL material. Additionally, the sensitivity to optical feedback is supported in many SL devices in which the facet reflectivity is only defined by the refractive index of the SL material. Since, the refractive index typically amounts to approximately $n \approx 3.5$, the facet reflectivity corresponds to about 30 %. This moderate reflectivity is sufficient for lasing operation which is due to the high gain of the SL material. However, this moderate reflectivity is also responsible for the low Q-factor of the SL resonator which, in turn, gives rise to the high susceptibility of SLs to perturbations, such as delayed optical feedback, or internal noise. In this context, it is necessary to know that SLs exhibit considerable spontaneous emission which represents a fundamental noise component entering the dynamical properties. This noise component can be included in the SL rate-equation model by adding Gaussian white noise, in form of Langevine noise terms, to the equation for the electric field, Eq. 2.9.

The collection of these excellent properties of SLs explains the interest in the utilization of this particular type of laser for studying generic NLD phenomena. Additionally, these qualities offers potential for optimization of the emission properties according to the technical demands. For both reasons, it is required to get a profound insight into the specific properties of the emission dynamics of SLs. In this context, understanding of the physical meaning of the α-parameter, representing the nonlinearity in the system, is required, since this parameter is of extraordinary importance to the emission dynamics of SLs. Furthermore, the properties of the relaxation oscillations of the carrier inversion in SLs are very crucial for the dynamical properties as well. This is due to the fact that the relaxation oscillation frequency commonly represents the upper limit of the bandwidth of the dynamics, while the damping rate of the relaxation oscillations affects the response characteristics of the dynamics to external perturbations. In the following, we discuss the fundamentals and properties of both peculiarities, the α-parameter and the relaxation oscillations, in detail. Additionally, we present experimentally obtained results for both characteristics, which give insight into their dependencies on other relevant system parameters.

The α-Parameter

SLs exhibit a particular strong amplitude-phase coupling of the electrical field in the laser medium. This means that a small change in the intensity (which can be induced by a spontaneous emission event, a transient in the injection current, or by external optical feedback) causes an excess perturbation of the phase of the lasing mode. The coupling between the amplitude and the phase is due to the physical nature of the gain medium of SLs. This difference with respect to other lasers originates from the particularities of the lasing transition in SLs, which in SLs occurs between two energy bands which are partially filled according to the Fermi-Dirac statistics, rather than between discrete energy levels. This property leads to an asymmetry of the gain curve [50] which, when Hilbert transformed, leads to a dispersion curve for the refractive index with zero crossing at higher frequency than the maximum of the gain curve [51]. Reason for this is that the optical gain and refractive index are connected via the Kramers-Kronig relations [52, 53], i.e., a change in the gain of the SL, induced by a variation of the carrier density, is accompanied by a change in the refractive index of the semiconductor material, and, hence, changes the phase of the lasing mode. Since the gain and the refractive index of the semiconductor material are directly related to the imaginary part, χ_i, and the real part, χ_r, of the susceptibility of the semiconductor material, $\chi(n) = \chi_r(n) + i\chi_i(n)$, the *Linewidth Enhancement Factor*, or briefly referred to as α-*Parameter*, can be expressed as follows [54, 55]:

$$\alpha = -\frac{d(\chi_r(n))/dn}{d(\chi_i(n))/dn} \qquad (2.11)$$

In contrast to SLs in which lasing results from inter-band transitions in most of the other types of lasers lasing transitions occur between discrete energy levels. The gain for discrete transitions is symmetric and, additionally, the symmetry does not depend on the inversion [38]. Therefore, the maximum of the gain curve and the zero position of the refractive index coincide and, hence, the α-parameter equals zero. This is the case for level-two lasers[1], a class to which many gas lasers belong.

Recently, estimation of the α-parameter for newly developed semiconductor laser structures, such as Quantum Dot Lasers [56] and Quantum Cascade Lasers [57] has attracted much interest. This interest arises from the fact that the lasing transitions are fundamentally different to common quantum well SLs. In Quantum Dot Lasers, lasing ideally occurs from transition between discrete energy levels; while in Quantum Cascade Lasers lasing is based on discrete intersubband transitions. Despite these differences, both new types of SLs are theoretically expected to exhibit zero α-parameter because of symmetric gain profiles. However, experiments have demonstrated α-parameters significantly differing from these first expectations which is due to different physical reasons. A non-zero α-parameter in Quantum Dot Lasers can be assigned to, e.g., inhomogeneous broadening due to the size distribution of the dots; while device self-heating effects represent one possible mechanism that can cause deviations from a zero-valued α-parameter for Quantum Cascade Lasers.

In the following studies, we concentrate on conventional bulk or quantum well edge emitting SLs, simply referred to as SLs. Generically, α-parameters of Quantum Well

---

[1]Two-level lasers for which the cavity resonance frequency is detuned from the frequency of the gain maximum can also exhibit a non-zero α-parameter.

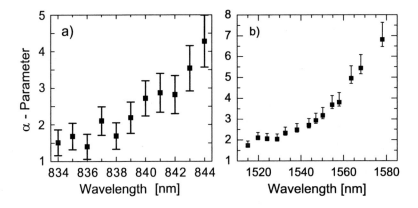

Fig. 2.2: Spectral dependence of the linewidth enhancement factor, the α-parameter, for a Hitachi HLP1400 Fabry-Perot and an IPAG DFB semiconductor laser, depicted in panels a) and b), respectively. The lasers emit at a center wavelength of 840 nm and 1541 nm, respectively. Results in panel a) with courtesy of T. Heil.

SLs significantly differ from zero, with typical values in the range between 1 and 7. Therefore, the α-parameter is one of the fundamental parameters determining the emission properties of SLs. In addition to the fact that α induces an excess linewidth broadening by a factor of $1 + \alpha^2$ [55, 58], the α-parameter plays a key role in many dynamical processes, e.g., in modulation induced frequency chirp [48], and, in particular, in delayed optical feedback induced dynamics [27]. For this reason, detailed knowledge of the α-parameter of the utilized SL is desired in many applications and dynamics experiments. Therefore, it is necessary to experimentally measure the specific α-parameter and its dependencies on other relevant system parameters. In the last three decades, many complementary methods have been developed for estimation of the α-parameter [59]. The most popular methods are based on measurement of the net gain of the SL material [60, 61], investigation of injection locking effects [62], or self-mixing interference phenomena [63].

Typical characteristics of the α-parameter are illustrated in Figure 2.2. The figure presents experimentally obtained results for the α-parameter of two different SLs in dependence of the emission wavelength. These two SLs are utilized for the experiments presented in Chapters 3 and 5. Figure 2.2 a) depicts results for a Fabry-Perot type SL (Hitachi HLP1400), emitting at a center emission wavelength of 840 nm, while the results obtained for a telecommunications distributed feedback (DFB) SL (IPAG DFB

SL, which is depicted in Figure 2.1), emitting at a center wavelength of 1551 nm, are shown in Figure 2.2 b). In the experiments, we have determined α according to the method proposed by Henning and Collins [61]. In this case, for both lasers we find an α-parameter of $\alpha \approx 2.5$ for the center emission wavelength. We note that this agreement is rather coincidence, since the material compositions of the lasers are different which determine the corresponding susceptibilities, and, hence, the corresponding α-parameters. The experimentally obtained α represents an effective α-parameter which depends on material specific properties and on the operation conditions of the SL. As an example, it has been shown that the device specific effective α can be considerably influenced by the structure of the laser resonator [64] or by thermal effects. These mechanisms can compensate for differences in the material susceptibilities so that the effective α-parameters of two different SLs can coincide, as it is the case here. However, the measured α-parameters represent typical values for conventionally available SLs. Furthermore, it becomes clear from the figure that both types of SLs exhibit similar spectral dependence of the α-parameter. For wavelengths shorter than the center wavelength, we find lower values for α, while the values for α increases for increasingly longer wavelengths. This spectral dependence is characteristic for the α-parameter and is in good agreement with theoretical predictions [65]. The spectral dependence of α can be linearized in vicinity of the solitary laser threshold:

$$\alpha = -2k\frac{d\mu/dn}{dg/dn} \tag{2.12}$$

Here, $d\mu/dn$ and dg/dn represent the derivatives of the refractive index μ and the gain per unit length g with respect to the carrier density n [59]; while k denotes the free space wave vector. It has been shown that the main contribution to the spectral dependence of the α-parameter can be assigned to the pronounced spectral dependence of the gain per unit length dg which is distinctly larger than the spectral dependence of the refractive index. Thus, the strong increase of α for increasing wavelength, which is highlighted by Figure 2.2, mainly originates from the strong decrease of the differential gain $d\mu$.

However, in most experiments it is not necessary to consider the spectral dependence of α, since the spectral dynamics is usually confined to a small range in which α does not vary significantly. Correspondingly, the α-parameter is considered as a constant in modeling. However, under certain experimental conditions this assumption is not valid, e.g., for spectrally filtered optical feedback with considerable detuning between feedback and SL emission, or in SL systems which exhibit exceptionally pronounced spectral dynamics. In Chapter 4, we study the emission properties of a SL system which exhibits such pronounced spectral dynamics. Now, we focus on the second relevant mechanism related to the dynamical properties of SLs, the relaxation oscillations.

The Relaxation Oscillation Frequency

The *Relaxation Oscillation Frequency* in SL represents a typical characteristic of Class B lasers. Relaxation oscillations can emerge in Class B lasers because they represent systems with (at least) two degrees of freedom. Such dynamically two-dimensional systems allow for periodic solutions. In SLs a periodic solution can emerge as oscillation of energy in the laser cavity between the carrier inversion and the optical field [48]. The corresponding frequency is defined by the characteristics of the light-matter interactions in the laser cavity:

$$\omega_{RO} = \sqrt{\frac{v_g \xi \overline{N_p}}{\tau_p}} \qquad (2.13)$$

This natural frequency is commonly referred to as the relaxation oscillation frequency, ω_{RO}. It is directly proportional to the square root of the differential gain, ξ, the group velocity in the gain medium, v_g, and the average photon density in the cavity, $\overline{N_p}$ (output power), and it is inversely proportional to the square root of the photon lifetime in the cavity, τ_p. Depending on the specific type of SL and the pump parameter, conventional SLs exhibit relaxation oscillation frequencies on sub-nanosecond time scale which is exceptionally fast if compared to other Class B lasers, such as Micro Disk Lasers, for which typical relaxation oscillation frequencies are several orders of magnitude slower.

For illustration of this particularity of SLs, Figure 2.3 presents the relaxation oscillations of the HLP1400 Fabry-Perot type SL recorded with a Streak Camera. The figure depicts the intensity response to a sudden turn-on event of the injection current to $p = 1.6\,I_{th,sol}$ of the SL revealing relaxation oscillations with a frequency corresponding to approximately 3 GHz. Furthermore, Figure 2.3 demonstrates that the oscillations are damped which is characteristic for solitary SLs. Nevertheless, this property can dramatically change if the SL is subject to perturbations. Then the relaxation oscillations can become of fundamental importance to the dynamical behavior of the SL, because even small perturbations can potentially undamp the relaxation oscillations. In particular, even small amounts of delayed optical feedback can perspicuously undamp the relaxation oscillations.

The high sensitivity of the emission dynamics of SLs to small perturbations can be used for determination of the frequency-current characteristics of the relaxation oscillations. Figure 2.4 depicts the results obtained for such measurements for the three types of lasers we utilize in this thesis.[2] In the previous subsection, we already introduced

---

[2]We provide a detailed overview over the characteristics of the different utilized SLs in Appendix A.1.

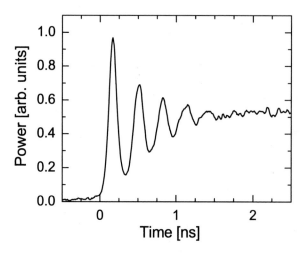

Fig. 2.3: Relaxation oscillations of a solitary HLP1400 semiconductor laser for sudden turn-
on process.

the HLP1400 Fabry-Perot type SL and the IPAG DFB type SL. Additionally, we
introduce a second Fabry-Perot type SL (FBH) emitting at a center wavelength of
786 nm. In the experiments, we apply a very small amount of optical feedback of
less than 0.1 % of the emitted power to each of the SLs. Then, we measure the peak
frequency of the undamped relaxation oscillations of the intensity dynamics of the
SLs. This procedure is repeated for different injection currents to determine frequency-
current characteristics, presented in Figure 2.4. The gray lines in the figure correspond
to a square root relation between the relaxation oscillation frequency and the injection
current. The experiment reveals that all of the SLs follow the theoretically predicted
square root behavior for pump parameters sufficiently above the solitary laser threshold
current $I_{th,sol}$. Only for pump parameters near $I_{th,sol}$, we identify some deviations from
the theoretically predicted square root law. These deviations can be caused by physical
spontaneous emission effects which are not considered in the model and on which the
lasers most sensitively reacts to near threshold. Furthermore, we find small deviations
from the predicted scaling of ν_{RO} for high levels of pumping, which can be attributed to
gain saturation effects, which have been neglected in the derivation of ν_{RO}. Figure 2.4
reveals that the IPAG SL exhibits the fastest relaxation oscillations of the lasers with
a maximum of approximately 7 GHz.

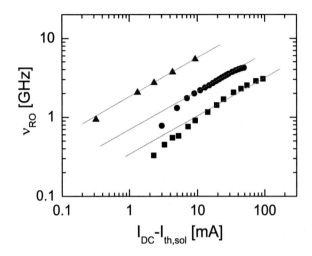

Fig. 2.4: Dependence of the relaxation oscillation frequency (ν_{RO}) of the utilized lasers on the pump current. Data obtained for different lasers is represented as follows: ▲ - IPAG, ● - HLP1400, and ■ - FBH. The gray lines follow the square root scaling that is predicted by theory for solitary semiconductor lasers.

Fast relaxation oscillations are strongly desired for telecommunication applications. Depending on the specific type of SL and the pump parameter, modern telecommunication SL can exhibit relaxation oscillation frequencies of up to several 10 GHz. This feature has boosted development of modern optical telecommunications applications based on SLs, since such high relaxation oscillation frequencies allow for sub-ns modulation of the intensity output of SLs which is required for today's multi GBit/s communication systems. In addition to the application-oriented point of view, the fast relaxation oscillations of SLs reveal their high potential for experimentally studying generic NLD phenomena, since they allow for extremely broadband dynamics comprising bandwidths of up to tens of GHz, roughly limited by the relaxation oscillation frequency.

The combination of both outstanding features of SLs, the high α-parameter and the fast relaxation oscillations, gives substantial reason for the distinguished role of SLs for experimental NLD. We will discuss this linkage between SL and NLD in the following subsection concluding this chapter.

2.3 Experimental Nonlinear Dynamics with Semiconductor Lasers

In the previous section, we have studied characteristic properties of the nonlinear interactions between the electric field and the gain medium in the laser cavity. We have identified the α-parameter and the relaxation oscillations as being the major characteristics determining the dynamical properties of SLs. On the one hand, the pronounced nonlinearity in SL, expressed by the α-parameter, provides an excellent basis for studying nonlinear effects. On the other hand, the extraordinary fast relaxation oscillations of SLs which can reach frequencies of up to several 10 GHz, depending on the pump parameter, offer potential for ultra broadband dynamics. Since these properties originate from device specific characteristics, such as material and cavity structure, the linkage between these advantageous properties is preserved for compound systems which are based on SLs. Therefore the distinguished features of SLs can be utilized for generation of high-dimensional broadband nonlinear dynamics, if additional degrees of freedom are added to the solitary SLs.

Today, high-dimensional nonlinear systems still reveal interesting unsolved phenomena. Many of these are beyond the scope of analytical treatment for reasons of complexity. Therefore, well-controllable experimental model-systems are desired to gain insight into generic properties of high-dimensional nonlinear dynamics. In this context, SLs suffice relevant premises for realization of model-systems that can capture high-dimensional NLD phenomena. In addition to their excellent nonlinear dynamical properties, SLs offer many practical advantages for experimental NLD. With respect to accurate experiments, it is of great value that SL exhibit well-accessible and well-controllable operation parameters, e.g., the pump parameter and the device temperature. Additionally, the devices are small and can be easily implemented into compound systems to increase the number of degrees of freedom. In this context, the high susceptibility of SLs to external perturbations is very complaisant, since only small perturbations are required for considerably increasing the number of degrees of freedom. This goal can be achieved by diverse approaches. A very simple, but efficient, approach is based on application of delayed optical feedback, in which the delay mathematically introduces infinite degrees of freedom to the SL system. Hence, SLs with delayed optical feedback are ideal devices for studying high-dimensional temporal NLD. Even more, the great variety of existing SL structures with different resonator design also allow for investigations of spatiotemporal nonlinear dynamics: conventional narrow-stripe SL represent transversally zero-dimensional[3] systems that are well-suited for study-

---

[3]In a strict sense only narrow stripe SLs which emit in the fundamental transverse mode and in which propagation effects can be neglected represent zero-dimensional SL systems. SL systems

ing temporal chaos; while modern high power Broad-Area Lasers (BALs) represent transversally one-dimensional systems that can exhibit complex spatiotemporal dynamics [66]; furthermore, transversally two-dimensional structures are provided by broad-area Vertical-Cavity Surface-Emitting Lasers (VCSELs), for which chaotic spatiotemporal dynamics [67] and recently wave-chaos have been revealed [68].

To summarize, SLs represent well-controllable and versatile model-systems for experimental NLD. Because of their excellent dynamical properties, in the last two decades SL systems have become established model-systems for NLD. Experimental NLD utilizing SLs has considerably contributed to today's understanding of generic nonlinear dynamical phenomena, and in particular has given insight into the emergence of chaos [23–27]. However, with respect to experimental and analytical treatability, so far, studies have mainly concentrated on low-dimensional NLD phenomena or on weakly pronounced high-dimensional chaotic emission dynamics of SLs. From the practical point of view, in turn, the gained knowledge of the onset of nonlinear dynamics in SLs is of great value, since it allows, e.g., for specific application of NLD for suppression of unwanted instabilities [69, 70].

Meanwhile, considerable progress has been made in NLD, SL physics, and measurement technology providing new tools for investigation of general properties of high-dimensional NLD phenomena. Anew, SLs with feedback can be utilized for this task, because they allow for controlled generation of various high-dimensional NLD phenomena, including high-dimensional broadband chaotic emission dynamics. Detailed characterization and analysis of the corresponding emission dynamical properties is desired. Furthermore, understanding of the influence of the systems parameters on the dynamical properties allows for tailoring of the emission properties according to application requirements. In contrast to previous intentions which aimed for suppression of chaos, novel applications have been proposed which are based on chaotic signals, e.g., encrypted communication [33] and high-resolution ranging applications [71]. However, generation of suitable chaotic signal is still an prevailing problem. In this context, SLs with delayed optical feedback represent very promising candidates to achieve this goal, since, as we will demonstrate in this thesis, they allow for generation of pronounced high-dimensional chaos with well-defined properties.

As a starting point, in the following Chapter 3 we extensively study the emission properties of SLs which are subject to delayed optical feedback. In particular, we focus on investigation and characterization of high-dimensional broadband chaotic emission dynamics, in which we are interested in terms of fundamental NLD and with regard to novel functionalities.

---

which are spatially extended in one ore two orthogonal directions which are transverse to the emission direction are denoted as transversally one ore two dimensional SL systems.

Chapter 3

Chaotic Dynamics of Semiconductor Lasers with Delayed Optical Feedback

In the late 1970s, it was reported that the emission properties of semiconductor lasers (SLs) can considerably change if the SL is subject to external optical feedback (FB) from a distant reflector [19]. Since then, SLs with FB have been intensively studied and motivated by two major reasons. On the one hand, the high sensitivity of SLs to FB is very important for *technical applications* of SLs. In many applications, SLs are unintentionally subject to reflections from distant system components which can easily destabilize the laser and result in complex emission dynamics [72–74]. This effect can impair functionality of various devices which are based on SL technology, e.g., in CD/DVD data storage/readout systems or in optical fiber based high-speed telecommunication devices.

As an example, Figure 3.1 illustrates the drastic effect of delayed FB on the emission properties of a conventional SL for telecommunications applications. In this situation, a SL (IPAG DFB SL) is solely coupled to a standard optical fiber by focusing the emitted light on the facet of the fiber placed at a distance of approximately 60 cm. Almost marginal FB from the facet of the fiber causes severe instabilities in the intensity output of the laser, which was stable for unperturbed solitary emission. Figure 3.1 a) discloses fast, sub-ns time scale, irregular pulsations in the intensity dynamics of the laser which represent characteristic dynamics for SLs with FB. The corresponding rf spectrum, which is shown in Figure 3.1 b), reveals the potentially high bandwidth of SL systems and characteristic peaks which originate from the delay in the system.

Usually, it is necessary to prevent such instabilities to guarantee functionality of the devices. One possibility to avoid FB induced instabilities is to introduce optical isolators for shielding FB. However, the implementation of optical isolators is associated with additional costs and contrary to often desired device miniaturization. Therefore, in recent years, the dynamics of experimental model-systems has been intensively studied to uncover the fundamental mechanisms leading to FB induced instabilities. From the applications point of view, insight into the emergence of nonlinear dynamics in

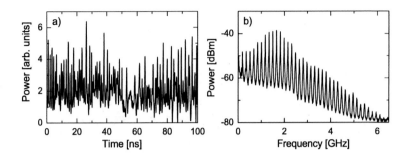

Fig. 3.1: Destabilized emission of a DFB semiconductor laser that is coupled to a distant
fiber and, hence, subject to delayed optical feedback from the facet of the fiber.
Panel a) demonstrates fast irregular intensity dynamics, while the corresponding rf
spectrum, presented in panel b), reveals the high bandwidth of the dynamics.

SLs allows for application of NLD for controlled suppression of feedback induced instabilities. Within the last three decades, sophisticated solutions have been developed
utilizing delayed feedback for stabilization of the output of SLs and for narrowing the
linewidth of the emitted light [70, 75–77].

In contrast to the commonly desired stabilization of the emission dynamics of SLs,
modern technology even aims for utilization of feedback induced instabilities. Recently,
the intriguing properties of high-dimensional chaotic dynamics have attracted much
attention since they offer potential for realization of novel applications which are based
on utilization of chaotic signals. In particular, SLs with FB offer auspicious dynamical
properties which can be harnessed for realization of innovative optical technology, e.g.,
for encrypted communication devices [78–82], for optical data readout systems and
frequency tuning [83], or for high-resolution chaotic LIDAR systems [71].

From the NLD dynamics point of view, SLs with FB are well-controllable nonlinear
systems. Therefore, in recent years, the emission dynamics of such SL systems has
been intensively studied and various nonlinear dynamical phenomena, comprising high-
dimensional chaotic intensity dynamics, has been identified [23–27]. The appearance of
different dynamical regimes in these SL delay systems essentially depends on the pump
parameter and on the main feedback parameters; the strength of the optical feedback;
i.e., the amount of light that reenters the active region of the SL, the length of the
external cavity (EC) and the feedback phase [81, 84–86]. Many of the results obtained
so far have given an insight into key questions of the field of Nonlinear Dynamics
(NLD) highlighting the high value of these practical experimental systems. Therefore,

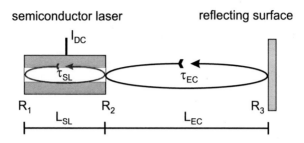

Fig. 3.2: Scheme of a semiconductor laser subject to delayed optical feedback.

SL systems with FB are well-suited for analysis of *fundamental NLD* phenomena of delay systems. In particular, utilization of SL with FB and application of state-of-the-art measurement technology allows for investigation of generic properties of high-dimensional nonlinear dynamics which has been an overly challenging task, so far.

This twofold motivation for studying SLs with FB underlines that research in this scientific field is extraordinarily relevant and exciting, because insight into fundamental NLD phenomena directly opens up perspectives for practical applications. Therefore, systematic experiments on the emission properties of SLs with FB in practicable experimental configurations are required. During the last years it has been shown that *External Cavity* configurations are very practicable and versatile configurations for studying feedback induced dynamics in SLs. Figure 3.2 schematically illustrates an external cavity SL (ECSL). In this configuration a solitary SL with facet reflectivities R_1 and R_2 is pumped by a constant injection current I_{DC}, while the light emitted by the SL propagates towards a distant mirror, with reflectivity R_3, from which it is reflected back into the SL after a time delay τ_{EC}. Hence, the SL is subject to time-delayed optical feedback. In the figure L_{SL} and L_{EC} denote the length of the SL and the external cavity (EC), respectively. Here the external cavity is defined by the front facet of the SL and the external mirror. Corresponding to both lengths, one additional relevant time scale is present in the ECSL system, in addition to the time scales of the light-matter interactions in the SL which we have discussed in Chapter 2. For this reason, ECSLs exhibit a large hierarchy of time scales which enter the dynamical properties. The most relevant time scales are the following three:

- The time corresponding to one round trip of the light inside the SL cavity: For conventional SL with a cavity length of $L_{SL} \approx 300\,\mu\text{m}$ the round trip time amounts to $\tau_{SL} \approx 7.2$ ps, which corresponds to a round trip frequency of $\nu_{SL} \approx 140$ GHz.

- The time corresponding to one round trip of the light in the EC, which defines the delay time in the system: The external cavity round trip time τ_{EC} depends on the length of the EC (L_{EC}) and the speed of light c: $\tau_{EC} = \frac{2L_{EC}}{c}$. In the experiments τ_{EC} can be substantially varied, typically between 50 ps and 1μs. The corresponding frequencies cover a range between $\nu_{EC} = 20$ GHz and $\nu_{EC} = 1$ MHz.

- The relaxation oscillation frequency ν_{RO}, which we have discussed in the preceding Chapter 2: ν_{RO} can reach frequencies of up to several 10 GHz. The relaxation oscillation frequency is of high importance for the dynamics, since it roughly defines the maximum bandwidth of the dynamics. Only under certain conditions, the bandwidth of the dynamics can substantially exceed ν_{RO}, which we will show later. Therefore, it is worth to know that this relevant frequency can be well adjusted by controlling the pump parameter, as we have demonstrated in Figure 2.4.

The delay time τ_{EC} is of particular importance for the dynamics of ECSLs. Depending on the delay time, ECSLs can operate in two qualitatively different regimes, the *Long Cavity Regime* (LCR), and the *Short Cavity Regime* (SCR) [49]. This denotation follows from the fact that the dynamics of ECSLs exhibits different characteristics for short and for long delay times, corresponding to short and long external cavities, respectively. A rough condition for categorization into these two regimes is given by comparison of the EC round trip frequency ν_{EC} and the relaxation oscillation frequency ν_{RO} [86, 87]. For situations in which the system fulfills $\nu_{EC} \gtrsim \nu_{RO}$ the system usually exhibits SCR dynamics, while the system operates in the LCR if the relevant frequencies suffice $\nu_{EC} < \nu_{RO}$.

In this chapter, we intensively study the emission properties of ECSLs operating in both regimes, the LCR and the SCR. In particular, we concentrate on investigation and characterization of the emission properties of ECSLs which exhibit high-dimensional chaotic emission dynamics. We study the influence of the relevant parameters on the emission dynamics applying methods of NLD. Additionally to the experimental analysis, we present complementary results obtained from numerical modeling to gain deeper insight into physical processes underlying the dynamics.

In the first section of Chapter 3, we present the experimental setup utilized for analysis of the dynamical properties of chaotically emitting ECSLs. We discuss typical dynamical properties of ECSLs depending the type of utilized SL, the applied feedback strength, and the operation conditions. In many cases, the dynamics of ECSLs can be modeled by the well-known Lang-Kobayashi semiconductor laser rate-equations allowing for complementary insight to the experiments into the dynamical properties. We recapitulate the assumptions underlying the Lang-Kobayashi SL rate-equation and discuss the model in terms of applicability to real-world ECSLs. In this context, we point out differences between singlemode and multimode lasers, both of which we use in the experiments. Subsequently, we experimentally investigate the intensity and spectral dynamics of chaotically emitting ECSLs and analyze its dependence on relevant system parameters. The experimentally obtained results are complemented by numerical modeling providing further insight into fundamental properties of the high-dimensional chaotic dynamics of ECSLs, which is experimentally not accessible. Finally, we compare characteristic features of chaotic dynamics generated in the LCR and in the SCR. The results will serve as basis for subsequent studies in which we propose two functional applications for high-dimensional chaotic dynamics generated by ECSL systems.

3.1 Characteristics of Feedback Dynamics: Experimental Requirements

At first glance, external cavity semiconductor laser systems represent very simple systems since they essentially consist of two sub-components, a SL and a mirror. However, these systems offer many well-accessible and well-controllable parameters which are relevant for their dynamical properties and which qualify ECSLs as versatile experimental model-systems for detailed studies on various NLD phenomena. In the following, we introduce a versatile experimental setup that allows for extensive characterization of the emission properties of ECSLs. Furthermore, we present characteristic dynamics of ECSLs for conventional operation conditions giving insight into the high demands on the detection apparatus for appropriately capturing the dynamics.

3.1.1 Experimental Setup

A scheme of the experimental setup of an ECSL system is depicted in Figure 3.3. The SL in the setup is pumped by an ultra-low-noise DC-current source and temperature stabilized with an accuracy of better than 0.01 K. Stabilization is required to guarantee constant environmental conditions, while minimization of external noise is necessary to obviate noise induced influences on the dynamics which can become relevant because of the high α-parameter of SLs. The light emitted by the SL is collimated by a lens (L) and propagates to a partially reflective mirror (PM) from which a fraction of the

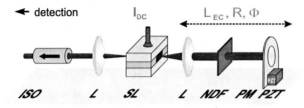

Fig. 3.3: Experimental setup for studying the dynamics of semiconductor lasers with delayed optical feedback.

light is reflected back towards the SL. The light reenters the laser after the time of flight, being $\tau_{EC} = 2L_{EC}/c$, with the phase (difference) $\Delta\Phi = \Phi(t) - \Phi(t - \tau_{EC})$. Here, L_{EC} stands for the length of the EC and c for the speed of light in air. The ratio between the power of the emitted light and that of the light effectively coupled back into the laser defines the strength of the feedback which is referred to as feedback ratio $r = \frac{P_{fb,eff}}{P_{out}}$. Such an ECSL system represents a delay system being mathematically infinite dimensional. The experimentally controllable key parameters of the system which determine the dynamical behavior are the delay time τ_{EC}, the feedback ratio r, the pump current I_{DC}, and the feedback phase $\Delta\Phi$. These parameters can be varied by changing L_{EC}, by replacing the neutral density filter (NDF), by varying I_{DC} and by shifting the PM on sub-wavelength scale with a piezo-electric transducer (PZT), respectively. The light emitted from the rear facet of the laser is sent to the detection branch which is isolated (ISO) from the system to prevent unwanted back-reflections.

In the detection branch, one part of the light is guided to an avalanche photodetector with a bandwidth of $\Delta\nu_{max} = 12\,\text{GHz}$ (New Focus 1544-A) for detection of the intensity dynamics. For further analysis of the dynamics, the output of the detector is monitored by a digital storage oscilloscope (DSO, Tektronix TDS7404) with an analog bandwidth of $\Delta\nu_{max} = 4\,\text{GHz}$ and a sampling rate of 20 GSamples/s. The spectral properties of the intensity dynamics are captured by an electrical spectrum analyzer (ESA, Tektronix 2755AP) with a bandwidth of $\Delta\nu_{max} = 21\,\text{GHz}$. Even higher temporal resolution can be achieved by application of a *Streak Camera* with a maximum detection bandwidth of $\Delta\nu_{max} > 50\,\text{GHz}$. Another part of the light is guided to an optical spectrum analyzer (OSA, Anritsu MS9710C) for investigation of the spectral properties of the ECSL system. The resolution of the OSA amounts to 50 pm, which is sufficient to resolve the longitudinal modes of SLs and to disclose moderate changes in the linewidth which might be induced by dynamics. These detection devices allow for comprehensive characterization of the emission dynamics of ECSLs in dependence on the relevant system parameters. So far, the studies of the dynamics of ECSLs have focused on the LCR, since the dynamics of long cavity ECSL systems is experimentally more easily accessible than that of short cavity ECSL systems. These studies have already revealed many interesting dynamical phenomena including chaotic dynamics. However, detailed studies of the emission properties for dynamics well within the chaotic regimes have not been performed so far.

3.1.2 Categorization of Long Cavity Regime Dynamics

In 1980 Lang and Kobayashi presented first systematic investigations of the emission properties of ECSLs in dependence on the feedback rate r and the EC length L_{EC} [88]. The measurements were performed in the conventionally studied long external cavity regime, with L_{EC} being typically longer than 10 cm. Lang and Kobayashi were able to describe the experimental results with a phenomenological model which is based on SL rate-equations. Five years later, Lenstra et al. demonstrated that the linewidth of a SL can increase by up to three orders of magnitude if the SL is subject to optical feedback. For that reason, this regime is referred to as *Coherence Collapse* regime (CC) [72]. Within a few years, the influence of r and p on the intensity dynamics of ECSLs has been intensively studied and categorized into five regimes [25,27,84]. In this scheme, introduced by Tkach and Chraplyvy [84], the CC regime is referred to as regime IV. The CC regime, in turn, has been further classified into five subcategories [89,90]. This was possible due to improvements in detection technology, which provided higher spectral and temporal resolution revealing new dynamical phenomena. Meanwhile, the influence of accessible system parameters on the dynamics of LC ECSLs operating in the CC regime is well-understood, i.e., temperature, pump parameter [89,90], feedback ratio [89, 90], tilt of the external mirror [91], and the external cavity length [85, 86]. Furthermore, effects of system specific parameters have been studied in depth. In this context, the most relevant parameters are the α-parameter [89, 90], the reflectivity of the laser facets [90], and the number of longitudinal modes being involved in the dynamics [92, 93]. Additionally, effects of specific types of feedback, such as phase conjugated feedback [94,95] and polarization-rotated optical feedback have been investigated [96,97]. However, detailed characterization of the dynamical properties of ECSLs which operate well beyond the onset of chaotic emission properties, in the so-called fully developed CC, is still lacking.

In this thesis we concentrate on experimental investigations of pronounced chaotic dynamics within regime IV, which can be achieved for moderate feedback ratios in a range typically between $0.01 < r < 0.1$. In regime IV, we observe the characteristic feedback induced instabilities in the laser intensity. Depending on the feedback ratio and the pump parameter, the stable emission of SLs can be easily destabilized leading to periodic, quasi-periodic or chaotic regimes of the intensity dynamics [84]. In particular, the CC regime reveals interesting emission properties comprising pronounced chaotic intensity dynamics in conjunction with chaotic spectral dynamics. Such dynamics manifests itself in broadening of the linewidth which is an indication of a substantial degeneration of the coherence properties of the laser light. An intriguing phenomenon of LCR chaotic dynamics in the CC regime is represented by the so-called *Low Frequency Fluctuations* (LFF). Characteristic for LFF dynamics are sudden, random dropouts of the laser intensity followed by a stepwise recovery process [25, 98, 99].

To give insight into fundamental properties and relevant time scales of the well-known LFF dynamics, in Figure 3.4 we present experimentally obtained results for LFF dynamics which is observed in the CC regime. In the experiment, we utilize a commercial Hitachi HLP1400 Fabry-Perot type SL. Details about the characteristics of the SL are provided in Appendix A.1. The laser operates near its solitary laser threshold $I_{th,sol}$, at $p = 1.08$, and it is subject to optical feedback with a time delay of $\tau_{EC} = 2.9$ ns and a feedback ratio of $r = 0.01$. The panels a) - c) depict segments of time series of the intensity dynamics acquired with different detection bandwidths. In Figure 3.4 a), the dynamics is captured by an oscilloscope with an analog detection bandwidth of 1 GHz. The figure reveals characteristic irregular dropouts of the intensity dynamics on a time scale of around one hundred nanoseconds. This time scale is slow if compared to characteristic time scales of the dynamics of the ECSL system, even if compared to the delay time which $\tau_{EC} = 2.9$ ns represents the slowest one. Increasing the temporal resolution by using an oscilloscope with an analog bandwidth of 4 GHz, we identify faster dynamics underlying the envelope of the characteristic low frequent (LFF) dynamics. This becomes clear when a Streak Camera is used for detection of the intensity dynamics which offers a detection bandwidth of exceeding 50 GHz. This high temporal resolution allows for fully resolving the intensity dynamics revealing underlying fast, irregular pulsations on picoseconds time scale. The corresponding optical spectrum is depicted in Figure 3.4 d). The gray line in this panel corresponds to the optical spectrum of the ECSL, while the black line represents the optical spectrum of the solitary laser in absence of feedback. Comparison of both spectra reveals the characteristic influence of FB on the spectral emission properties of the laser. For solitary emission the side modes of the Fabry-Perot SL are suppressed by approximately 20 dB with respect to the central longitudinal mode located at 840.6 nm. This property drastically changes when FB is applied to the SL. In this case, we observe a characteristic redshift ($\Delta\lambda$) and linewidth broadening for the central longitudinal mode of the SL. Furthermore, the same effects on the spectral properties are observed for the other longitudinal modes which are excited because of the FB and contribute to the intensity dynamics.

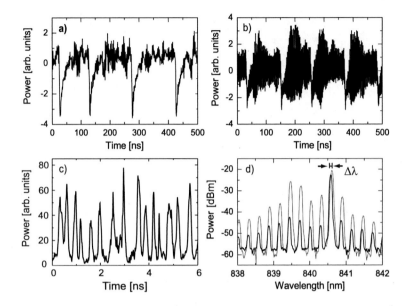

Fig. 3.4: Example of feedback induced instabilities of a Fabry-Perot semiconductor laser (Hitachi HLP1400). Panels a) - c) present intensity time series for similar operation conditions which have been recorded with different detection bandwidth, $\Delta\nu_{max}$: in panel a) $\Delta\nu_{max} = 1\,\mathrm{GHz}$, in b) $\Delta\nu_{max} = 4\,\mathrm{GHz}$ and in c) $\Delta\nu_{max} > 50\,\mathrm{GHz}$. Corresponding spectral emission properties with (gray line) and without feedback (black line) are represented by the optical spectra depicted in panel d). Comparison of the spectra reveals redshift and linewidth broadening for feedback induced dynamics which is characteristic of dynamics in the coherence collapse regime.

This property raises the question whether the characteristic LFF dynamics is influenced, or even induced, by competition effects between the longitudinal modes. If that were the case, the dynamics would represent unique characteristics of Fabry-Perot lasers exhibiting weak side mode suppression. To answer this question, we investigate the emission properties of a second ECSL system. In contrast to the previous experiment using a Fabry-Perot type SL, we now implement a singlemode DFB SL in the ECSL setup with a side mode suppression of 40 dB. We perform a comparable experiment choosing the following conditions: $p = 1.09$, $r = 0.05$, $\tau_{EC} = 2.9\,\mathrm{ns}$. The results are presented in Figure 3.5. Panel a) depicts the corresponding intensity time series which

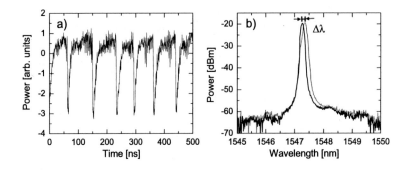

Fig. 3.5: Example of feedback induced instabilities of a DFB SL. Panel a) presents intensity time series for similar operation conditions as in Figure 3.4 which have been
recorded with a detection bandwidth of $\Delta\nu_{max} = 1$ GHz. The corresponding optical
spectrum is depicted in panel b).

has been captured with the 1 GHz bandwidth oscilloscope. The time series is in good
qualitative agreement with that of Figure 3.4 a). We note that for the singlemode DFB
laser, we find similar characteristic irregular pulses on fast sub-ns time scale, which is
indicated by the fast fluctuation on top of the LFF envelope in the time series in panel
a). The corresponding optical spectrum is shown in panel b). Here, the black solid line
represents the optical spectrum of the solitary laser, while the gray solid line depicts
the optical spectrum in presence of optical feedback. Both spectra expose the good side
mode suppression ratio of the DFB laser of about 40 dB. In similarity to Figure 3.4 d),
Figure 3.5 b) reveals a distinct redshift ($\Delta\lambda$) of the emission line accompanied by a
substantially broadening of the linewidth. However, in this case the side mode suppression is strong enough to prevent FB induced excitation of additional longitudinal
modes.

In summary, we find well-agreeing characteristics of the dynamics of both Fabry-Perot
lasers and of DFB SLs. Only one exception originates from the weak side mode suppression properties of Fabry-Perot lasers for which we find multimode dynamics, while
singlemode emission persists for the DFB laser. Several other FB induced characteristics of the intensity dynamics and of the spectral dynamics can be observed for
both different types of lasers. From these results, we can conclude that under these
conditions the influence of multimode effects can be neglected. We note that in recent years the role of multimode effects for LFF dynamics, such as competition for
the common gain, has been an intensively studied question and controversially de-

bated [100, 101]. Today, various experiments have verified that multimode effects are of minor importance under many feedback conditions and that Fabry-Perot SLs and DFB SLs essentially exhibit similar dynamics features [49, 90]. However, under certain conditions multimode effects can become important and cannot be neglected [93], i.e., in presence of strong feedback or in SL systems with very small longitudinal mode spacing [102]. Most of the experiments we present in the following consider investigations for conditions in which multimode effects are negligible. Only one exception is presented in Chapter 4, in which we consider a unique ECSL system for which multimode effects are crucial.

In conclusion, ECSLs which operate in the LFF regime under typical conditions, i.e., moderate feedback and moderate pump parameters, exhibit pronounced intensity and spectral dynamics comprising a multitude of time scales and a distinct spectral bandwidth, respectively. The corresponding enfolding bandwidth of the dynamics emphasizes the high demands on measurement technology with respect to appropriate analysis of the complex dynamics of ECSLs. Only recently developed measurement technology provides reasonable temporal and spectral resolution to allow for appropriate characterization of such complex dynamics. Since profound experimental characterization of ECSLs is very challenging, complementary modeling of the dynamics is desirable which can provide further insight into the mechanisms underlying the dynamics of ECSLs. Numerical modeling can complement experimental characterization of the dynamics, in particular with respect to studies on high-dimensional chaotic NLD phenomena with even higher dynamical bandwidth and with respect to applications thereof. For this reason, in the next section we present a simple modeling approach based on the SL rate-equations, which can provide further insight into fascinating properties of the dynamics of ECSLs.

3.2 Modeling Feedback-Induced Dynamics of Semiconductor Lasers

So far, we have demonstrated that ECSLs can exhibit irregular intensity dynamics, comprising a broad range of time scales, in conjunction with considerable spectral dynamics. These dynamical properties are very interesting from the fundamental NLD point of view. Detailed experimental analysis of ECSLs which can give insight into the fundamental properties of the dynamics is possible, but it requires sophisticated measurement technology. Experimentally we only can gain insight into properties of the dynamics utilizing accessible parameters and variables. In real-world experiments, however, usually some of the relevant system variables and parameters are difficult to measure or are even inaccessible. In particular, in ECSLs this applies to gain properties and phase of the complex electric field. In this context appropriate modeling can provide further insight into fundamental processes underlying the dynamics. This is due to the inherent property of models that the included parameters and variables are principally, mathematically accessible. Of course, the major difficulty of modeling consists in developing an appropriate model that accurately mimics the experiment. – But what is the conceptional meaning of "appropriate" in terms of modeling?

As the expression "model" already implies, mathematical modeling refers to a theoretical construct which is devised to mimic processes of real-world systems. Therefore, mathematical models necessarily represent subjective and therefore simplified representations of real world systems. From the physical point of view, modeling serves as a means to enhance the scientists insight into the real-world system by uncovering experimentally hidden processes. In that sense, a model can be treated as a good model if it not only reproduces observed phenomena, but even more so if it provides new insight into inaccessible processes and if it allows for predictions of new physical phenomena. Predictions, in turn, are obliged to be categorically verifiable. Conversely, a mathematical model needs to be abandoned or refined if its predictions are in contradiction to experimental observations.

In the next subsection, we will follow these considerations and present a simple mathematical model for description of the emission properties of ECSLs. We will demonstrate that although the model is based on several, to some extent even severe, assumptions its results and predictions are in excellent agreement with experimentally observed phenomena of conventional ECSLs. The model provides insight into fundamental processes leading to the dynamical instabilities of ECSLs and their characteristic emission properties.

3.2.1   Lang-Kobayashi Semiconductor Laser Rate-Equations

In Chapter 2, we have discussed the SL rate-equation model 2.9 - 2.10 which has become an established model for description of the emission properties of solitary SLs. The SL rate-equations can serve as basis for modeling ECSLs. To describe the dynamics of ECSLs, one has to account for delayed optical feedback in the model. In 1980 Lang and Kobayashi proposed an extended version of the SL rate-equations [88]. In their model, they introduced an additional expression $\gamma e^{-i(\Phi(t)+\Phi_0)} E(t - \tau_{EC})$ to the equation of the complex electric field, Eq. 2.9, incorporating delayed optical feedback. The resulting rate-equations are commonly referred to as *Lang-Kobayashi* equations:

$$\dot{E}(t) = \frac{1}{2}\big(1 + i\alpha\big)\xi n(t)E(t) + \gamma e^{-i(\Phi(t)+\Phi_0)}E(t - \tau_{EC}) \tag{3.1}$$

$$\dot{n}(t) = (p-1)\frac{I_{th,sol}}{e} - \frac{n(t)}{T_N} - \big(\Gamma + \xi n(t)\big)|E(t)|^2 \tag{3.2}$$

The first term on the right-hand side of the field equation Eq. 3.1 attributes to the solitary laser emission. The second term accounts for the external optical feedback, including the external cavity round trip time τ_{EC}, and the feedback rate γ, respectively. Multiple round trips of the light in the external cavity are neglected. The phase accumulated by the electric field within one round trip in the external cavity is $\Phi_0 = \omega_0\tau_{EC}$ mod 2π, where ω_0 represents the optical frequency of the solitary laser. In the equations the variables are expressed with respect to the threshold conditions of the solitary laser. Therefore, $n(t)$ represents the excess carrier number (density) with respect to the solitary laser threshold level: $n(t) = N(t) - N_{th,sol}$. The other parameters are: the α-parameter, α; the differential gain, ξ; the solitary laser threshold current, $I_{th,sol}$; the pump parameter, $p = \frac{I_{DC}}{I_{th,sol}}$; the electron charge, e; the carrier lifetime, $T_N = 1/\gamma_N$; and the cavity decay rate, Γ.

The LK rate-equations, Eqs. 3.1 - 3.2, represent a dissipative, deterministic model for ECSLs. However, in the LK-equations the implementation of delayed optical feedback is based on several assumptions. The major ones comprise neglect of multiple reflections of the light in the EC and restriction to coherent feedback fields. Furthermore, this basic model does not consider noise effects which, under certain conditions, might become important because of the high sensitivity of SLs even to small perturbations, which we pointed out in Chapter 2. Additionally, nonlinear gain saturation effects are not taken into account which potentially can influence the dynamics for moderately strong feedback and high injection currents. Both effects, noise and nonlinear gain saturation, can be included in the model, if required. In this case it is convenient to describe the dynamics of the ECSL in terms of the physically relevant absolute values of the output variables, which allows for direct comparison of experimentally obtained results with results obtained from modeling. On the other hand, the corresponding notation better reveals the influence of gain saturation effects. Accordingly, the rate-equations governing the electric field and the carrier number read:

$$\dot{E}(t) = \frac{1}{2}\big(1 + i\alpha\big)\big[G(t) - \Gamma\big]E(t) + \gamma e^{-i(\Phi(t)+\Phi_0)}E(t - \tau_{EC}) +$$
$$\sqrt{2\beta\gamma_N N(t)}\,\zeta(t) \tag{3.3}$$

$$\dot{N}(t) = \frac{p\,I_{th,sol}}{e} - \frac{N(t)}{T_N} - G(t)|E(t)|^2 \tag{3.4}$$

$$G(t) = \frac{\xi\big(N(t) - N_0\big)}{1 + \varepsilon|E(t)|^2} \tag{3.5}$$

Symbol	Meaning	Value
α	linewidth enhancement factor	4
ξ	differential gain	$6 \times 10^{-6}\,\text{ns}^{-1}$
Γ	cavity decay rate	$200\,\text{ns}^{-1}$
T_N	carrier life time	$0.5\,\text{ns}$
N_0	transparent carrier number	1.5×10^8
ϵ	gain suppression coefficient	5×10^{-7}
β	spontaneous emission factor	10^{-7}
$I_{th,sol}$	solitary threshold current	≈ 60 mA

Table 3.1: Physical and device parameters for modeling.

In the model, complex Gaussian random terms $\zeta(t)$ are added in the field equations to model spontaneous emission processes. These numbers have zero mean, $\langle \zeta(t) \rangle = 0$, and correlation, $\langle \zeta(t)\zeta^*(t') \rangle = 2\delta(t - t')$. The gain function in Eq. 3.5 is approximated to depend linearly on the instantaneous carrier number and accounts for gain-suppression effects. Meaning of the remaining parameters can be found in Table 3.1. In order to give an impression of reasonable values for numerical modeling, we also present device-typical values for the Hitachi HLP1400 SL. Depending on the particular experiment, we utilize both versions of the LK-equations for complementary analysis of the experiments. On the one hand, analytical treatment of the LK-equations can give insight into local and global structures of the phase space in which the dynamics evolves. However, analytical treatment is challenging because the LK-equations are differential delay equations which are mathematically infinite dimensional. Additional difficulties arise from the broad range of time scales involved in the dynamics. Therefore, analytical insight into the dynamical properties of ECSLs is challenging. Multiple time scale analysis techniques [103] are promising to achieve such insight. Nevertheless, such analytical treatment of the LK-equations is far from being simple so that most of the insight into the dynamical properties of the LK-equations is currently restricted to basic solutions, such as determination of fixed points and periodic solutions with its corresponding stability analysis. On the other hand, numerical modeling allows for application of methods of time series analysis. Once the model parameters are experimentally determined and the model agrees with the experiment, the equations can be numerically integrated with appropriate step size. Then, numerically obtained time series allow for quantitative estimation of information-theory based measures, such as *Lyapunov Exponents, Information Dimension, Kolmogorov-Sinai* entropy. This is very useful, since these measures are experimentally not accessible because of the presence of noise and presently existing limits in detection bandwidth and resolution. Therefore, numerical modeling can be beneficial if the following two preconditions are fulfilled to guarantee meaningful comparison of experiments and numerics. Firstly, the system parameters need to be estimated with sufficient accuracy. Secondly, applicability of the model must be justified in terms of the assumptions on which the model is based.

3.2.2 Suitability of the Model

Derivation of the LK-model is based on several assumptions. Therefore, a priori it is not evident that the model-system indeed captures the major physical properties of the real experiment. In the following, we study suitability of the LK-model with respect to its application to real experiments. Therefore, at first, we check if the model can reproduce typical dynamical phenomena exhibited by the experimental system. Then, we concentrate on the question if the model provides useful insight into general properties of the dynamics of ECSLs which are experimentally inaccessible.

We chose LFF dynamics to serve as touchstone for testing the suitability of the LK-model. For modeling we adopt typical experimental conditions for which we obtain LFF dynamics and substitute the corresponding parameters into the LK rate-equations, Eqs. 3.1-3.2. The conditions for modeling roughly correspond to the experiment presented in Figure 3.4. We consider an ECSL which consists of a HLP1400 SL with moderate feedback ratio, $r = 0.05$ corresponding to $\gamma = 22\,\mathrm{ns}^{-1}$; low pump parameter, $p = 1.04$; and an external cavity round trip time of $\tau_{EC} = 2.9\,\mathrm{ns}$. For modeling the properties of the SL, we insert the experimentally obtained parameters for the HLP1400 SL, presented in Table 3.1, into the LK-equations. With this set of parameters we numerically integrate the LK-equations to mimic the dynamics of the ECSL. The corresponding intensity time series are depicted in Figure 3.6 a). The figure clearly reveals typical characteristics of LFF dynamics. In agreement with the experimental results presented in Figure 3.4, on the one hand we find irregular intensity dropouts occurring on a slow time scale of the order of 100 ns. On the other hand, a zoom into the time series exposes fast, irregular intensity fluctuations on several 10 ps time scale. This behavior is illustrated in Figure 3.6 b), in which we present the intensity dynamics of one LFF event of Figure 3.6 a). In agreement with the experimental findings, the model reflects the development of the dynamics during one LFF event in which the short-time-averaged intensity slowly builds up, fluctuates around a certain level, until it eventually, suddenly drops to zero-level, followed by the subsequent recovery phase of the next LFF event. The good agreement between the intensity dynamics obtained by modeling and experiment is particularly remarkable, since the HLP1400 SL does not remain longitudinal singlemode in presence of FB like it is assumed in the LK-

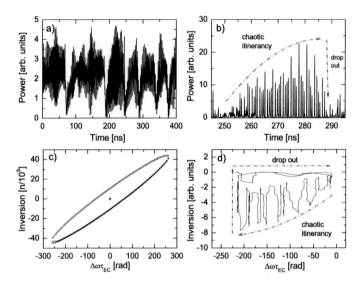

Fig. 3.6: LFF dynamics modeled with the Lang-Kobayashi SL rate-equations. Panel a)
depicts 1 GHz low-pass filtered intensity dynamics. A zoom of one LFF scenario of
panel a) is presented in panel b) in which the bandwidth corresponds to 27 GHz.
The time series reveals fast pulsations on sub-ns time scale. The corresponding
fixed point structure (ellipse) in the inversion-frequency space is depicted in panel
c). The filled circles denote modes, the open circles mark anti-modes, and the
diamonds at the bottom of the ellipse represent stable modes. The star indicates the
solitary laser mode. Panel d) illustrates the dynamics of the LFF cycle presented
in b) in inversion-frequency space. The parameters are: $\tau_{EC} = 2.9$ ns, $p = 1.04$,
and $\gamma = 22$ ns^{-1}.

model. Nevertheless, the intensity dynamics of both, modeling and experiment, reveal
the same characteristic features being typical for LFF dynamics. Based on the good
agreement between modeling and experiment, we study solutions to the equations to
understand the characteristic evolution of the dynamics during one LFF event.

To attain fundamental information about the dynamical properties of the LK-model,
it is straightforward to analyze the fixed point structure of the deterministic model
equations, Eqs. 3.1 - 3.2. The fixed point solutions can be obtained by substituting
the ansatz $E(t) = Ae^{i\omega t}$, $n(t) = n = \text{const}$ into Eqs. 3.1 - 3.2 leading to a set of

transcendental equations, Eqs. 3.6 and 3.7 [25, 104]:

$$\Delta\omega\tau_{EC} = \gamma\tau_{EC}\sqrt{1+\alpha^2}\sin\left[(\omega_0 + \Delta\omega)\tau_{EC} + \arctan\alpha\right] \tag{3.6}$$

$$(\gamma\tau_{EC})^2 = \left(\Delta\omega\tau_{EC} - \frac{\alpha}{2}n\xi\tau_{EC}\right)^2 + \left(\frac{n\xi\tau_{EC}}{2}\right)^2 \tag{3.7}$$

The solutions to this problem reveal a fixed point structure located on an ellipse around the solitary laser mode in the phase space spanned by the inversion, n, and the frequency, ω [98]. In this space, each of the fixed points is defined by a constant optical frequency $\omega = \omega_0 + \Delta\omega$ and a constant carrier number n, where ω_0 denotes the frequency of the solitary laser mode. Figure 3.6 c) depicts the fixed points corresponding to the time series in Figures 3.6 a) and b). On this ellipse, the fixed points with higher gain are shifted to lower frequencies if compared to the solitary laser mode. This property originates from the α-parameter. The fixed points on the ellipse occur pairwise while the two fixed points of each pair have different stability properties. These two solutions are referred to as *Mode* and *Anti-Mode* [105]. In the figure the modes and anti-modes are represented by filled and open circles, respectively. The solitary laser mode is located in the center of the ellipse and represented by a star. For increasing the feedback rate γ, the modes and anti-modes consecutively are created in pairs at the low frequency end of the ellipse. This occurs in a *Saddle-Node* bifurcation from which the modes follow the stable branch, while the anti-modes follow the unstable branch. Subsequent to the saddle-node bifurcation, each newly created mode may destabilize in a *Hopf* bifurcation for increasing feedback rate. However, at least one mode at the low-inversion edge of the ellipse is always stable. Since for this mode inversion is lowest, it exhibits highest gain, appointing its name *Maximum Gain Mode* (MGM). Depending on the α-parameter, additional modes in the vicinity of the MGM can remain stable, which are referred to as *High Gain Modes* (HGMs). The HGMs meet the following requirement [49, 106] and are denoted as diamonds in Figure 3.6 c):

$$-\arctan\alpha^{-1} < (\omega_0 + \Delta\omega)\tau_{EC} < 0 \tag{3.8}$$

Physically, the modes represent solutions corresponding to constructive interference condition between the feedback field and the field in the SL, while the anti-modes correspond to the destructive interference condition. For this reason, the fixed point solutions are also referred to as *External Cavity Modes*. The ellipticity, the orientation and the extension of the ellipse in the $(\Delta\omega, n)$-space and, additionally, the number of external cavity modes influence the dynamical properties [49]. Analysis of Eqs. 3.6 and 3.7 discloses that the eccentricity of the ellipse is determined by α, while the extension of the ellipse depends on α and γ.

$$\pm\Delta\omega_{max} = \pm\alpha\gamma \tag{3.9}$$

The number of modes between the solitary laser mode and the MGM, Z_{modes}, can be determined from Eq. 3.9 by division of the frequency spacing of two adjacent external cavity modes, i and j, $|\Delta\nu_{i-j}| = 2\pi\tau_{EC}^{-1}$.

$$Z_{modes} = \frac{\alpha}{2\pi}\gamma\tau_{EC} \qquad (3.10)$$

Based on these static solutions, we are able to phenomenologically explain the LFF phenomenon in terms of dynamics states evolving along the destabilized fixed point solutions. Figure 3.6 d) represents the intensity dynamics of the LFF event of Figure 3.6 b) in frequency-inversion space. Comparison of the intensity dynamics and the dynamics presented in frequency-inversion space disclose that the intensity dynamics is directly related to spectral dynamics. According to the LK-model, the LFF dynamics can be understood as chaotic itinerancy of the dynamics along the attractor ruins of the (Hopf) destabilized modes [104, 107]. After a dropout of the intensity the dynamics state itinerates towards higher intensity and lower frequency. In this process, the laser erratically emits intense pulses on ps time scale. Each pulse in the intensity time series can be attributed to one fast change of the inversion in vicinity of a mode in the frequency-inversion space. In the short periods of zero intensity between two pulses the dynamics can jump from one attractor ruin to another neighboring one. Since gain increases for decreasing frequency, the dynamics state preferably jumps to modes located at lower frequencies. In this way the dynamics continuously approaches the mode with lowest inversion, represented by the MGM, which is accompanied by steadily increasing intensity. In vicinity of the MGM, the stable manifold approaches the unstable manifold of the anti-modes [70] resulting in a crisis. Therefore, the dynamics state can eventually hit the unstable manifold so that it is repelled from the high gain region. Then, the intensity suddenly drops, the SL stops lasing for an instant, and the dynamics state is reinjected close to the solitary laser mode. Finally, this process continuously repeats itself in succeeding LFF events, that is why it has been named *Sisyphus Effect* [98].

We highlight that the numerically obtained results, presented in Figure 3.6, are obtained from the LK-equations without additional noise terms in the equations. This means that the major characteristics of the experimentally observed LFF dynamics, comprising fast irregular ps time scale pulsing dynamics and low frequency intensity dropouts, can be understood by application of a purely deterministic model. This gives evidence for a deterministic nature of the instabilities in ECSLs. However, we note that noise can influence the statistic properties of the chaotic dynamics [108, 109]. Additionally, we point out that the LK-model does not represent the only deterministic model which mimics the characteristic features of the LFF dynamics of ECSLs. In Chapter 2 and 3, we have seen that the LK-model is based on several assumptions.

In particular, the simplifications regarding the description of the light-matter interaction in the laser cavity and the restriction to singlemode emission do not necessarily apply to all ECSLs. Therefore, alternative sophisticated models have been proposed to account for detailed descriptions of these simplified processes [101, 110–116]. Some of these models represent extended models which are based on the LK rate-equations. Other more complex models are based on partial differential equations that account for spatial dependencies in the SL cavity.

However, the presented example of LFF dynamics demonstrates that the LK-model meets both requirements of a good model we are looking for. Firstly, the qualitative good agreement between the LFF dynamics obtained form the LK-model and the experiment confirms that the model captures relevant physical mechanisms underlying the dynamics. Secondly, the LK-model facilitates complementary insight into the dynamics and allows for disclosing of fundamental physical mechanisms and understanding of dynamical phenomena of ECSLs. This has also been verified for modeling of numerous other experimentally observed dynamical phenomena [25, 49, 86, 87, 98, 107, 117–119]. Therefore, this simple, phenomenological model has proven its suitability for modeling and analysis of generic properties of the diverse nonlinear dynamics of ECSLs.

The great advantage of the LK-model originates from its balance between complexity and simplicity. The model is complex enough to capture numerous NLD phenomena, but it is just simple enough to allow for some elementary analytic investigations and to facilitate physical interpretation of the results. However, some peculiar ECSLs or conventional ECSLs under extreme operation conditions exhibit dynamical phenomena that are beyond capability of the LK-model. To prevent inapposite application of the LK-model its restrictions need to be considered with respect to the particular experimental conditions. In the following, we point out restrictions of the LK-model and briefly discuss alternative modeling approaches.

Restrictions of the LK Model – Alternatives

One relevant assumption for the LK-model consists in restriction to *single longitudinal mode* operation of the SLs. Modern DFB SLs with high side mode suppression ratio meet this requirement, but in contrast to this, conventional Fabry-Perot type SLs usually emit on several longitudinal modes (LMs) if they are subject to FB. However, comparison of modeling, Figure 3.6, and experiment, Figure 3.4, has revealed good agreement. Similar agreement has been verified in numerous experiments for different dynamical regimes, provided that only a small number of longitudinal modes is excited and coupling between adjacent longitudinal modes is weak. We note that the agreement between experimentally observed multimode dynamics and singlemode LK-modeling only holds for comparison of the total intensity dynamics, while the dynamics of the individual LMs can exhibit characteristic features originating from mode competition effects which are not captured in the basic LK-equations. To capture modal dynamics, extended versions of the LK-equations have been proposed accounting for lasing and coupling of several longitudinal modes [92, 101, 115, 116, 120–125]. In one class of these LK based models, each additional LM is considered by one additional field equation, while all LMs deplete the same shared carrier reservoir. In these models, the LMs modes are either coupled through self- and cross-interaction processes [101, 115] or a mode dependent gain is considered [121]. In the second class of LK multimode models, in spirit of the Tang-Statz-deMars [126] model, additionally to the extra field equations different carrier densities are assumed for each mode [92, 116]. In the past years, these diverse models have substantially contributed to observations and understanding of interesting multimode dynamical phenomena [127]. Each of the models exhibit different dynamical properties which have been recently compared by Koryukin et al. [125]. However, presently it is still an open question which of the models does capture multimode dynamics best, since many of the dynamics differences are subtle and all of the models have their pros and cons according to the underlying approximations. A detailed description of multimode effects can be established in the framework of sophisticated traveling wave models. Traveling wave models consider spatial effects occurring in ECSLs which are neglected in the LK-model. Such a detailed treatment of the emission properties of ECSLs necessitates sophisticated mathematics since the corresponding model consists of partial differential equations.

The LK-model represents a *Mean-Field Model* in which spatial extensions of SLs are disregarded. This neglect can become of relevance either for large SLs structures such as Broad Area Lasers, or for comparable extensions of the laser and the external resonator. In both cases, several physical effects can originate from the spatial extensions of the devices, which can be captured by *Traveling Wave* models. Firstly, in real SLs counter-propagating traveling waves occur in the SL cavity that are coupled to the carrier reservoir and, therefore, induce longitudinal mode dynamics [113]. Secondly, diffraction, carrier diffusion, and structural inhomogeneity of SLs can quantitatively and qualitatively modify the emission dynamics [112, 128]. Thirdly, possible resonance conditions between longitudinal modes and external cavity modes can potentially enhance coupling between adjacent longitudinal modes influencing the spectral dynamics. In Chapter 4, we will present such a system which exhibits pronounced spectral dynamical properties. In this context, traveling wave models have demonstrated their potential for modeling the dynamics of SL systems including spatial effects [111–114, 128, 129]. However, these models also have their limits of validity and they can therefore not be applied in general. Additionally, physical interpretation of the results obtained form traveling wave models can be challenging because of the required sophisticated mathematical handling.

Further approximations and assumptions of the LK-model might constrain its applicability in special cases. On the one hand, the basic LK-equations only consider spectral dynamics comprising *moderate spectral bandwidth*. This fact originates from linearization of relevant physical parameters which in reality exhibit nonlinear spectral dependence. For pronounced spectral dynamics, which can be induced by strong FB, these approximations can give rise to deviations between experiment and modeling. In this context, the linearly approximated gain and the constant α-parameter represent the most relevant parameters. On the other hand, the implementation of the optical feedback in the LK-model does not consider multiple reflections in the external cavity which can become of importance for SLs with high facet reflectivity that are subject to strong feedback. In this case, multiple delay times come into play, which can be implemented in a LK-based model. Finally, the basic LK-model represents a purely deterministic model. In contrast to this assumption, noise is inevitable in real world systems. In the experiment noise with different properties can originate form various physical and environmental sources effecting both, the electric field and the injection current. Conventionally, only spontaneous emission noise is considered in extended LK-models, such as in Eqs. 3.3-3.4. Spontaneous emission noise represents the strongest noise source in SLs for continuous emission, since it changes the electric field which induces linewidth broadening because of the moderate α-parameter. In contrast to this, carrier noise does not significantly influence the linewidth of SLs operating continuous wave and, therefore, it is conventionally neglected. However, we note that carrier noise terms can influence the stability properties of the dynamics which, in turn, determine the final state of the dynamics [130].

Major parts of the investigations we present in the following focus on temporal dynamics of conventional ECSLs for which it is warrantable to neglect spatial effects. We present experimentally observed dynamical phenomena which are in good agreement with results obtained from numerical modeling within the framework of LK justifying application of the LK modeling approach. Henceforward, we refer to these ECSL systems as *Lang-Kobayashi Systems*. Only one exceptional system will be studied in Chapter 4 which exhibits dynamical phenomena that are clearly beyond capability of the LK modeling approach. In the following, we extensively investigate the emission properties of pronounced chaotically emitting Lang-Kobayashi ECSLs. In particular, we characterize the dynamics utilizing methods of NLD and analyze the influence of relevant system parameters on the dynamical properties. Our goal is to gain insight into fundamental processes determining the dynamical properties and to utilize the obtained knowledge for controlled generation of high-dimensional broadband chaotic emission dynamics with well-defined properties. Tailoring of the emission dynamics of

SLs by harnessing NLD is desired with respect to novel applications which are based on chaotic signals.

3.3   Chaotic Dynamics of Lang-Kobayashi Systems: The Long Cavity Regime

During the last decade, analysis of the dynamics of ECSLs has mainly focused on EC-SLs operating in the long cavity regime (LCR) in which the external cavity round trip frequency $\nu_{EC} = 1/\tau_{EC}$ is sufficiently smaller than the relaxation oscillation frequency ν_{RO} of the SL. This is motivated by two reasons. On the one hand, experiments in the LCR are very robust and the essential system parameters are well-controllable. On the other hand, the dynamics in this regime reveals interesting NLD phenomena comprising also dynamics which exhibits conspicuous features on comparatively slow time scales that are experimentally well-accessible, e.g random switching between stable emission and dynamics, and irregular intensity dropouts in LFF dynamics. For these reasons, ECSLs operating in the LCR are excellently suited for studying the onset of chaotic dynamics in high-dimensional delay systems, especially when limitations in resolution and bandwidth of the detection devices are inevitable. Therefore, previous experimental work has mainly focused on investigation of the emergence of chaos and studies of generic phenomena within the LFF regime, while detailed analysis and characterization of the properties of chaotic dynamics beyond the LFF regime, in the fully developed *Coherence Collapse* (CC) regime, are still lacking.

In this section, we present a comprehensive characterization of the emission properties of chaotically emitting ECSLs operating in the LCR. In particular, we focus on dynamics generated in distinctive chaotic regimes. Starting and reference point for our investigation is the well-known LFF regime from which we proceed to fully developed, broadband chaotic dynamics which can be found in the CC regime. We experimentally investigate the dynamics in both regimes benefiting from complementarily performed numerical modeling. Therefore, we characterize the dynamical properties in dependence on relevant system parameters utilizing methods of NLD. The combination of both, the joint experimental and numerical approach, and the application of methods of NLD, facilitates deep insight into the dynamical properties of chaotically emitting ECSLs.

3.3.1 From Low Frequency Fluctuations to Fully Developed Coherence Collapse

Recently, application of high-resolution temporal and spectral detection devices allowed for first comprehensive experimental characterization of LCR dynamics in dependence on the relevant system parameters [49, 89, 90]. The presented results revealed different dynamical regimes for variation of the pump parameter, the feedback rate γ and the α-parameter. It has been realized that for further advancing into the chaotic regime the demands on the detection devices strongly increase. This is particularly the case for dynamics generated by ECSLs with moderate feedback for pump parameter well beyond the solitary laser threshold current $I_{th,sol}$, for which ECSLs operate well within the fully developed CC regime.

To provide a basis for forthcoming studies, we specify typical operation conditions for which we find regimes of pronounced chaotic dynamics. Figure 3.7 provides a global overview over the dynamical regimes of the HLP1400 SL operating in the LCR. The figure illustrates the influence of the key parameters of the system, the pump parameter p (represented in terms of the injection current I_{DC}) and the feedback rate γ, on the dynamics of the ECSL. The considered ranges for the parameters correspond to typical experimentally accessible values. We note that feedback rates between $40\,\mathrm{ns}^{-1} > \gamma > 10\,\mathrm{ns}^{-1}$ span the range of moderate feedback corresponding to effective feedback ratios between $0.01 < r < 0.1$. In the figure, we identify the four regimes of dynamics that have been discovered by Heil et al. [89], marked as shaded regions in the γ-I_{DC}-space. These four regimes comprise stable emission, coexistence of stable emission and LFF dynamics, LFF dynamics, and CC dynamics.

Each of these four regimes is passed through for increasing the injection current for moderate feedback conditions. As an example we discuss the development of the dynamics following the route marked by the arrow at $\gamma = 27\,\mathrm{ns}^{-1}$. For low pump currents in vicinity of the feedback reduced laser threshold, $I_{DC} \gtrsim I_{th,fb}$, we find stable emission on several external cavity modes (ECMs). For increasing I_{DC} the dynamics enters the coexistence regime in which the dynamics exhibits irregular transitions between constant emission and LFF dynamics. This regime is represented in different shades of moderately dark gray. The longer the stable emission state persists, the darker the shade of gray. Further increasing I_{DC} leads to the LFF regime which is represented by the region filled in light gray. In the LFF regime the dynamics does not reveal stable emission, but characteristic irregular intensity dropouts on low frequency time scale. The characteristic intensity dropouts vanish for higher I_{DC} when the dynamics emerges to the CC regime, the regime of fully developed chaotic dynamics.

Figure 3.7 reveals that the regimes of operation crucially depends on the chosen system

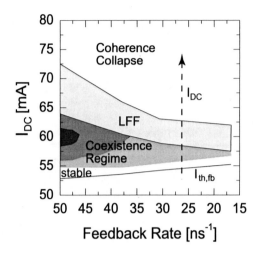

Fig. 3.7: Dynamical regimes of the HLP1400 SL operating in the long cavity regime in dependence on the injection current I_{DC} and feedback strength γ for $\alpha \approx 2.5$. By courtesy of T. Heil [49].

parameters. Additionally, we note that the transitions between the regimes are not abrupt indicating that the dynamical properties within the regimes significantly depend on the operation parameters. Since we are interested in studying well pronounced chaotic dynamics, we concentrate in the following on investigations within the LFF regime and the CC regime. Both regimes are predominately encountered for moderate feedback strength, which we need to account for in the experiments. For this condition, the desired dynamical regime can be adjusted by variation of the pump parameter. LFF dynamics dominates for low pump currents near the solitary laser threshold current, with pump parameters $p \approx 1$; while CC dynamics prevails for moderate pump currents, typically, with pump parameters $p > 1.1$.

3.3.2   Properties of Chaotic Long Cavity Regime Dynamics

In the previous subsection we have given a qualitative overview over the dynamics which can be observed in the LCR. In this subsection, we present a comprehensive analysis and characterization of the dynamical properties of ECSLs operating within the LFF and the CC regime. In particular, we study the influence of the relevant system parameters on the dynamical properties using the experimental setup described in Subsection 3.1.1. Our characterization will comprise analysis of the spectral and the auto-correlation properties of the intensity dynamics, experimental estimation of the dimension of the dynamics, investigation of the effect of applied perturbations on the dynamics, and influences of the intensity dynamics on the spectral emission properties. Furthermore, we present results for complementary performed numerical modeling allowing for determination of information theoretical ergodic measures of the dynamics such as Lyapunov exponents, Kaplan-Yorke dimension and Kolmogorov-Sinai entropy.

Influence of the Pump Parameter on the Intensity Dynamics

The pump parameter p represents a well-accessible control parameter for the dynamics of ECSLs which can be easily changed. Figure 3.7 has already indicated that the dynamical properties may considerably change for variation of p. The figure shows that by increasing the pump parameter for moderate feedback conditions, the dynamics evolves from the LFF regime to the CC regime. However, this qualitative categorization does not provide information about the actual effect of the pump parameter on the dynamics. To gain insight into the corresponding dynamical properties, we perform experiments using the HLP1400 SL in the experimental configuration described in Subsection 3.1.1.

In the experiment, the SL is temperature stabilized and subject to delayed optical feedback from an external cavity with $L_{EC} = 40\,\text{cm}$, corresponding to a delay of $\tau_{EC} = 2.7\,\text{ns}$. We utilize a neutral density filter to adjust for moderate feedback strength of $r = 0.05$ for which we measure a threshold reduction of $1 - I_{th,fb}/I_{th,sol} = 0.053$. For these conditions, the ECSL operates in the LCR in which we study the influence of the pump parameter p on the intensity dynamics of the SL. The intensity dynamics is detected by a fast photodetector with 12 GHz bandwidth which converts the light emitted by the SL into an electric signal. We monitor this rf signal using an oscilloscope with 4 GHz analog bandwidth and a rf spectrum analyzer with 21 GHz analog bandwidth.

For low pumping of $p = 1.08$, near the solitary laser threshold, the ECSL exhibits characteristic dynamics for the LFF regime. This is demonstrated in Figure 3.8 a), which depicts a 300 ns long segment of the intensity dynamics. The time series discloses three sudden irregular dropouts of the intensity. These dropouts are followed by a recovery phase of the short time average intensity lasting approximately 30 ns. This time roughly corresponds to $10\,\tau_{EC}$. With the onset of the recovery phase, we find fast irregular pulsing behavior of the intensity on sub-ns time scale. This pulsing behavior persists even when the short time average intensity has built up. The average

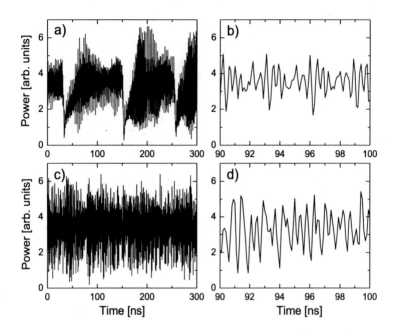

Fig. 3.8: Emission dynamics of a HLP1400 laser in the LFF regime at $p = 1.08$ depicted
in panels a) and b) and at $p = 1.30$ shown in panels c) and d). The dynamics is
measured with a photodetector and a fast digitizer with an analog bandwidth of 4
GHz. The laser is subject to moderate feedback of $r = 0.05$ and the length of the
external cavity is $L_{EC} = 40$ cm, which corresponds to $\tau_{EC} = 2.7$ ns. Panels b) and
d) depict 10 ns zooms of a) and b), respectively.

intensity remains almost constant after the built-up phase until it eventually drops
and the procedure starts anew. Figure 3.8 b) presents a 10 ns long zoom into the time
series shown in a) when the dynamics has finished the recovery process. The zoom
into the dynamics discloses irregular fast sub-ns pulsing of the intensity. Thus, the
dynamics on the fast and the slow time scales reveals all the typical characteristics of
LFF dynamics, we have previously discussed in detail. Additionally, from the zoom
into the intensity dynamics we can identify that the dynamics exhibits similarity with
its preceding state one delay time before. This similarity at shifts of $\tau_{EC} = 2.7$ ns
becomes evident when comparing the intensity dynamics at around 93.3 ns with the
dynamics at around 96.0 ns. Since the light emitted by the SL is reinjected into the

laser after τ_{EC}, the SL is "reminded" on its past state on which it reacts. This feature induces a memory effect which is characteristic for delay systems.

For higher pump parameters, the dynamical properties change. For increasing p, we find that the average frequency of the intensity dropouts increases until this low-frequency characteristic of the dynamics vanishes. Figure 3.8 c) illustrates intensity dynamics obtained for $p = 1.30$. For this moderate pump level, the intensity time series does not give rise to low frequent dropouts. Now, the short time average intensity of the dynamics is almost constant, while the average intensity level rises for increasing p. Nevertheless, the dynamics on the fast sub-ns time scale reveals fast, irregular pulsing behavior with high modulation depth. These are indications for dynamics of the fully developed coherence collapse regime. The corresponding dynamical properties are highlighted in Fig 3.8 d) in which we present a 10 ns long segment of the time series of Figure 3.8 c). When increasing the pump parameter, the comparison of Figures 3.8 b) and d) reveals that the pulses become faster and increasingly irregular. Furthermore, for dynamics within the CC regime, we do not identify obvious similarity of the dynamical states with its previous states one delay time before, as it is the case for LFF dynamics observed for low levels of pumping.

These observations indicate that the dynamical properties significantly depend on the pump parameter p. For the transition from LFF to CC dynamics, which we find for increasing p, analysis of the intensity time series reveals three main effects. Firstly, the low frequency component of the dynamics related to the intensity dropouts vanishes. Secondly, the sub-ns pulsing becomes faster. Thirdly, the dynamics exhibits increasingly irregular behavior for raising the pump parameter. The corresponding dynamical properties can be characterized by use of appropriate analysis techniques, such as spectral and correlation analysis.

Bandwidth of Dynamics

We have seen that the intensity time series of the emission properties of the ECSL exhibit significant dependence of the dynamics on the pump parameter. However, since segments of time series present snapshots of the dynamics, without further processing they provide only limited insight into the qualities of the dynamics. This is particularly the case for dynamics that exhibits irregular behavior and that comprises high spectral bandwidth. Then, spectral analysis of the time series can provide comprehensive insight into the dynamical properties, which can be done by calculation of the Fourier transform. However, sufficiently high temporal resolution and appropriately long time series are required to fully capture the dynamical properties in this way. Unfortunately, it is technically very challenging to realize oscilloscopes that offer sufficiently high sam-

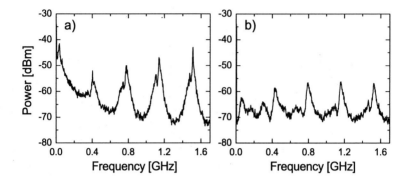

Fig. 3.9: Low frequency part of the rf spectra corresponding to the time series presented
in Figure 3.8. In panel a) the injection parameters amounts to $p = 1.08$, while
$p = 1.30$ in panel b).

pling rate and record length to fully capture the dynamics of chaotically emitting SLs,
which can exhibit dynamics comprising a dynamical bandwidth of up to several 10 GHz.
Such oscilloscopes are just being developed. Current state-of-the-art oscilloscopes offer
analog bandwidths of up to 15 GHz with sampling rates and memory depths of up
to 40 GSamples/s and 64 GSamples, respectively. Nevertheless, the spectral properties
of the intensity dynamics can also be directly measured by utilization of the hetero-
dyne technique. This sensitive measurement method is implemented in electrical (rf)
spectrum analyzers (ESA), allowing for analysis of the dynamics in frequency space
covering high spectral bandwidth. In this way, electrical spectrum analyzers can provide useful information about the (time-averaged) spectral properties of the dynamics
evading the very complex technology required for time series analysis.

Figure 3.9 depicts the low-frequency part of the rf spectra of the dynamics corresponding to the time series of Figure 3.8. In Figure 3.9 a), we present the spectrum for
LFF dynamics which we find for $p = 1.08$. The spectrum discloses broad peaks superimposed by characteristic low-frequency dynamics. This low-frequency dynamics
exhibits decreasing power for increasing frequencies. For the low-frequency dynamics,
we identify a peak at 31 MHz which corresponds to a period of 32 ns. This represents the most probable duration time of the LFF events, while the probability for the
occurrence of shorter LFF events drastically reduces until it vanishes for frequencies
exceeding 1.3 GHz. The characteristic broad peaks, on the other hand, are related to
faster dynamics. These peaks are equally spaced and exhibit a small dip near their

center frequencies. The dip of the first peak is located at a frequency of 375 MHz, which corresponds to the inverse of the delay time $\tau_{EC} = 2.7$ ns. Therefore, the peak is related to the delay, while the occurrence of harmonics of ν_{EC} can result from the nonlinearity of the system giving rise to inharmonic dynamics or it might due to the stability properties of the external cavity modes, determined by its respective eigenvalues [131]. We note that previous explanations for the observed splitting of the peaks have been proposed [132, 133], but the fundamental mechanism underlying this phenomenon is not understood in detail and subject to ongoing investigations. Our experimental investigations regarding this problem so far suggest that the splitting of the peak results from the interplay between delay effects and relaxation oscillations.

The broadband character of the spectrum gives strong evidence for chaotic behavior of the dynamics[1]. This property manifests itself in the time series in Figure 3.8 a) as the observed irregular pulsing and unpredictable sudden dropouts of the intensity. Since we have identified distinct differences for higher pump parameters, we are interested in how the spectral properties of the dynamics change when we increase the pump parameter to $p = 1.30$. The result is presented in Figure 3.9 b). For moderate pumping well beyond the threshold, the rf spectrum does not exhibit pronounced low-frequency components anymore. This finding agrees with our observations for the intensity time series, presented in Figure 3.8 b), which did not give evidence for the characteristic low-frequent intensity fluctuations of the LFF regime. Even more, we identify significant broadening of the peaks located at around ν_{EC} and multiples thereof. A closer look at the spectrum discloses that the positions of the dips within this peaks coincide with their positions for lower pumping, as it can be seen from Figure 3.9 a). Interestingly, the splitting of the peaks increases when increasing the pump parameter. In this process, the peaks considerably broaden and merge, so that it is difficult to relate them to the well-defined peaks observed for low pump levels. As a consequence thereof, the effective bandwidth of the dynamics increases.

So far we have focused on the low frequency part of the dynamics. To study the properties of the fast frequency components of the dynamics, we depict in Figure 3.10 the rf spectra of the same measurement, but now with an extended spectral range of up to 7 GHz. Both spectra reveal additional harmonics of the dynamics related to the cavity round trip frequency $\nu_{EC} = 375$ MHz. However, we find distinct differences for the envelopes of the peaks when comparing both cases. In Figure 3.10 a), the corresponding envelope exhibits a maximum at about 2.2 GHz. This frequency roughly equals the re-

---

[1]For comparable conditions, it has been shown in numerical modeling that the number of positive Lyapunov exponents can exceed 100 [134] giving rise to high-dimensional chaotic dynamics. We will address the question about dimensionality of the dynamics and other dynamics measures later on in this chapter, where we present numerical results providing further insight into the dynamical properties and their dependence on relevant system parameters.

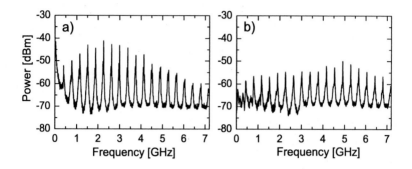

Fig. 3.10: Almost full range rf spectra of the dynamics for similar conditions as in Figure 3.8. In panel a) the injection parameters amounts to $p = 1.08$, while $p = 1.30$ in panel b).

laxation oscillation frequency of the solitary laser, which corresponds to $\nu_{RO} = 1.9\,\text{GHz}$ for $p = 1.08$. For higher frequencies, the envelope describes an exponential decay of roughly 40 dB per decade. In contrast to this, the envelope of the peaks in Figure 3.10 for $p = 1.30$ does not exhibit such a clear defined maximum, since the heights of the peaks only slightly vary within the presented range. However, we can identify a weakly pronounced maximum near 4.9 GHz. This frequency is considerably beyond the relaxation oscillation of the solitary laser for $p = 1.30$, being $\nu_{RO} = 2.9\,\text{GHz}$. For higher frequencies, the power of the peaks decreases similarly as in the case of low pumping.

To sum up, the spectral properties of the intensity dynamics disclose enhanced bandwidth for increasing pump parameters. For the transition from the LFF regime to the CC regime, two effects of enhancement can be observed. On the one hand, the maximum bandwidth of the dynamics raises due to the increasing relaxation oscillation frequency. Consequently, the dynamics involves faster frequency components. On the other hand, the peaks related to the delay dynamics significantly broaden giving rise to increasing irregularity of the dynamics. Nonetheless, the spectral properties of the dynamics do not provide sufficient information about the irregularity of the dynamics. The question about irregularity properties is directly related to the occurrence of correlations between temporally spaced states of the dynamics. Since the correlation properties of the dynamics are influenced by the nonlinearity of the system, analysis of the auto-correlation properties of the dynamics can provide further insight into nonlinear properties of the dynamics.

Auto-Correlation Properties

Auto-correlation analysis of the intensity dynamics can give insight into relevant properties of the dynamics, since it allows for identification of important time scales of
the dynamics and the related physical processes. Therefore, investigation of the influence of the system parameters on the correlation properties can provide first insight
into the development of the nonlinear behavior. Thus auto-correlation analysis represents a beneficial tool which is also easy to accomplish. Furthermore, given that
auto-correlation analysis is a statistical measure it even provides useful information for
moderately resolved experimental time series, for which sophisticated methods of time
series analysis may fail, e.g., determination of the Lyapunov spectrum and information
dimension.

To gain further insight into the dynamical properties of the ECSL system, we calculate
the auto-correlation functions of the measured intensity time series. For this procedure,
it is advantageous to normalize the time series to avoid considerable dc-contributions
entering the auto-correlation function, since it can hamper identification of relevant
features of the dynamics. Therefore, we shift the time series to zero mean power,
calculate the resulting variance of the power, and normalize the variance to unity.
Then, we calculate the auto-correlation function of the normalized intensity time series
as following:

$$\Sigma_{corr}(\Delta t) = \frac{\langle \delta P(t + \Delta t)\delta P(t)\rangle_t}{\sqrt{\langle \delta P^2(t + \Delta t)\rangle_t \langle \delta P^2(t)\rangle_t}}. \tag{3.11}$$

Here, Σ_{corr} represents the correlation coefficient, $\delta P(t)$ denotes the instantaneous normalized power at time t, while Δt stands for the time shift with respect to t. The angled
brackets indicate time averaging over t, which has been performed for time series with
at least a length of $500\,\tau_{EC}$. Figure 3.11 presents the calculated auto-correlation functions for the time series depicted in Figure 3.8. The auto-correlation function in Figure 3.11 a) corresponds to the dynamics within the LFF regime observed for $p = 1.08$.
Please note that the time is scaled in units of the delay time $\tau_{EC} = 2.7\,\text{ns}$. For low pump
parameters, the auto-correlation exhibits many sharp peaks with high correlation at
the delay time and multiples thereof. This can be seen in the inset of Figure 3.11 a). In
the vicinity of these peaks, at about plus-minus $0.2\,\text{ns}$, we identify considerable negative
values of the correlation coefficient giving evidence for anti-correlation. The correlation between these peaks decreases from approximately $\Sigma_{corr} = 0.25$ near $t \approx 0\,\tau_{EC}$
to $\Sigma_{corr} \approx 0$ for $t = 3\,\tau_{EC}$. The maxima of the peaks describe a symmetric envelope
which reveals recurrent increases of the correlation for time intervals corresponding
to 12 round trips of the light in the cavity. Since, $12\,\tau_{EC} = 32.4\,\text{ns}$ corresponds to
$\nu = 31\,\text{MHz}$, the frequency corresponding to this recurrent correlation agrees with the

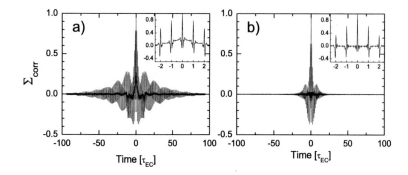

Fig. 3.11: Auto-correlation function of the time series of Figure 3.8 for $p = 1.08$ and $p = 1.30$
b). The time scale is normalized to the external cavity round trip time being
$\tau_{EC} = 2.7$ ns corresponding to $L_{EC} = 40$ cm.

characteristic frequency of the intensity dropouts, we have identified in Figure 3.9 a).
Furthermore, the envelope discloses a fall-off of the correlation to values below $1/e$ for
time intervals exceeding $2.8\,\tau_{EC}$ or 7.6 ns, respectively. The auto-correlation function
decreases to almost zero after approximately 100 round trips of the light in the cavity.
This means that there is no correlation between the state of the actually generated
dynamics and the state of the dynamics advancing or lagging by $t = 100\,\tau_{EC}$.

Next, we study the effect of increasing p on the auto-correlation properties. Figure 3.11 b) presents the auto-correlation function for $p = 1.30$. The figure shows a
similar shaped auto-correlation function, but now the auto-correlation exhibits distinctly faster fall-off if compared to Figure 3.11 a). In analogy to Figure 3.11 a), the
inset in Figure 3.11 b) also reveals high correlation for time intervals corresponding to
the delay time and multiples thereof. However, the correlation background between
these peaks has vanished for dynamics within the fully developed coherence collapse.
Hence, the actual state of the dynamics is only correlated to future or past states of
the dynamics at multiples of τ_{EC}. Furthermore, we find that the envelope of the auto-
correlation function for $p = 1.30$ rapidly drops to values below $1/e$ for time intervals
exceeding $t = 1.9\,\tau_{EC}$ which corresponds to $t = 5.1$ ns.

The comparison of the auto-correlation properties of the LFF dynamics and the CC
dynamics, presented in Figures 3.11 a) and b), respectively, reveals several characteristics.

- The intensity dynamics within the LFF and the CC regime is correlated to earlier (later) states of the dynamics at τ_{EC} or multiples thereof. This gives evidence for memory effects due to the delay, representing a characteristic property of delay systems.

- The auto-correlation of the dynamics at multiples of the delay time decreases for increasing p. This indicates, that the dynamics becomes increasingly irregular, faster diverging from similar dynamical states. Reason for this is the enhancement of the bandwidth of the dynamics, as it can be identified from the rf spectra in Figure 3.10.

- The envelope of the auto-correlation functions exhibit recurrent correlations on a time scale of several τ_{EC}. Since these long-range correlations are particularly pronounced for LFF dynamics, the recurrent correlations can be attributed to the characteristic intensity dropouts.

Our analysis of the auto-correlation properties of the dynamics highlights the key-role of the pump parameter for the dynamics. As a rule of thumb, for moderate feedback strength, the irregularity of the dynamics increases when increasing the pump parameter. Additionally, the results disclose that the delay time apparently plays a key role for the dynamics. Therefore, in the following, we study the influence of variations of τ_{EC} on the auto-correlation properties of the dynamics.

Without restrictions, we choose the CC regime for the corresponding experiments by setting the pump parameter to $p = 1.30$. To obtain comparable results to the previous experiments, we again adjust for a moderate feedback rate of $r = 0.05$. Then, we keep all parameters but L_{EC} constant and study the influence of τ_{EC} on the intensity dynamics. Therefore, we vary the external cavity length between $4\,\mathrm{cm} < L_{EC} < 330\,\mathrm{cm}$ and record the intensity time series from which we calculate the auto-correlation functions, as we have described before. The results are illustrated in Figure 3.12. The figure exemplarily depicts the auto-correlation function for four different lengths of L_{EC}. In panel a) $L_{EC} = 322\,\mathrm{cm}$, $\tau_{EC} = 21.5\,\mathrm{ns}$; in panel b) $L_{EC} = 80\,\mathrm{cm}$, $\tau_{EC} = 5.3\,\mathrm{ns}$; in panel c) $L_{EC} = 40\,\mathrm{cm}$, $\tau_{EC} = 2.7\,\mathrm{ns}$; and in panel d) $L_{EC} = 7\,\mathrm{cm}$, $\tau_{EC} = 0.5\,\mathrm{ns}$. Please note that the auto-correlation functions are presented in absolute nanoseconds time scale, while the time of the presented segments corresponds to $20\,\tau_{EC}$ for all of the auto-correlation functions. Interestingly, we find similarly shaped envelopes of the auto-correlation functions within the whole range of delay times. Only the absolute time corresponding to the range of the auto-correlation function depends considerably on the delay time. We find the following times $t_{e^{-1}}$ corresponding to a fall-off of the envelope of the auto-correlation to $1/e$: in a) $t_{e^{-1}} = 42\,\mathrm{ns}$, in b) $t_{e^{-1}} = 8.9\,\mathrm{ns}$, in c) $t_{e^{-1}} = 5.1\,\mathrm{ns}$, and in d) $t_{e^{-1}} = 0.9\,\mathrm{ns}$. The same values scaled in units of the delay

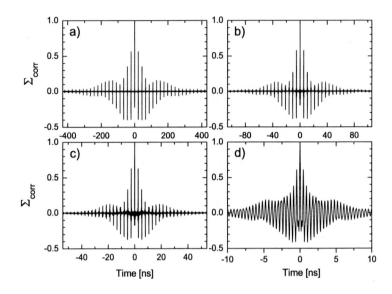

Fig. 3.12: Dependence of the auto-correlation properties of the intensity dynamics on the external cavity length (L_{EC}) for $p = 1.30$ and $r = 0.05$. The conditions are: $L_{EC} = 322$ cm in a), $L_{EC} = 80$ cm in b), $L_{EC} = 40$ cm in c), and $L_{EC} = 7$ cm in d). These lengths correspond to round trip times of $\tau_{EC} = 21.5$ ns, $\tau_{EC} = 5.3$ ns, $\tau_{EC} = 2.7$ ns, and $\tau_{EC} = 0.5$ ns.

time τ_{EC} are: in a) $t_{e^{-1}} = 1.9\,\tau_{EC}$, in b) $t_{e^{-1}} = 1.7\,\tau_{EC}$, in c) $t_{e^{-1}} = 1.9\,\tau_{EC}$, and in d) $t_{e^{-1}} = 1.8\,\tau_{EC}$. Since the decay interval is almost constant in units of the delay time τ_{EC}, the absolute time corresponding to the range of the auto-correlation linearly scales with τ_{EC}. This means that variation of the delay time within the long cavity regime only quantitatively changes the auto-correlation properties, while it does not significantly affect the qualitative auto-correlation properties of the dynamics.

From this result, we can conclude that the mechanism which introduces the correlations in the dynamics represents a general property of the delay, since it is independent of its absolute value. The mechanism seems to be related to the properties of the interference occurring between the dynamical state of the light emitted by the SL and the dynamical state of the feedback that reenters the SL at the same time and that represents the dynamical state of the system one delay time before. In this process the actual state of the dynamics approaches the previous state of the dynamics maintaining the correlation for the next consecutive interference event one delay time later. At the present time the actual mechanism underlying this process in ECSLs is not fully understood. Experimentally, it is very difficult to access this problem, since investigations require full access to dynamics. However, appropriate experimental determination of the states of the dynamics is actually not feasible. This is a particularly challenging task for the dynamics of ECSLs operating under the described conditions for which the dynamics is potentially high-dimensional. However, modeling approaches may provide further insight into the fundamental mechanisms determining the properties of the feedback induced correlations.

Nonetheless, we can utilize the experimentally obtained auto-correlation functions to extract further information about the dynamical properties. In this context, the auto-correlation functions can be used to roughly estimate the dimension of the dynamics. This idea is based on the *Le Berre Conjecture*. According to M. Le Berre "the Lyapunov dimension of chaotic attractors is almost equal to the delay time divided by the correlation time of the feedback driving force" [135]. The correlation time in the Le Berre conjecture is introduced as the relaxation time of a system that is required before the system can anew react on the feedback. For ECSLs, this time roughly corresponds to the time of the first zero-crossing of the auto-correlation function. Hence, we measure the intensity time series for various lengths of the external cavity L_{EC}. Then, for each of the time series we estimate the time $t_{0,\Sigma=0}$ of the first zero-crossing of the auto-correlation function. Finally, we divide each τ_{EC} by $t_{0,\Sigma=0}$ to obtain a rough estimate of the dimension of the dynamics in dependence on the delay time. Figure 3.13 represents the results we have achieved for the same feedback ratio of $r = 0.05$ as for the experiments presented before. We have estimated the dimension of the dynamics for two different pump parameters, $p = 1.08$ and $p = 1.30$. The data for $p = 1.08$ is represented by triangles in Figure 3.13, while the results for $p = 1.30$ are marked as asterisks, respectively.

The results disclose estimated dimensions for the dynamics in the range between 2 and 250. We can identify a linear scaling of the estimated dimension with the external cavity length and, thus, with the delay time: $\tau_{EC}/t_{0,\Sigma=0} \propto L_{EC}$. The observed linear growth of the dimension for increasing delay time is in good agreement with numerically obtained results for other delay systems, e.g., Mackey-Glass models for white-cell production, optical bistable hybrid systems, and a nonlinear ring cavities [135]. Linear scaling is observed for both pump parameters, while the dynamics in the CC regime for $p = 1.30$ gives evidence for higher dimensionality than for the dynamics in the LFF regime for $p = 1.08$. This suggests that the dimension of the dynamics grows for increasing p. Intuitively, this result is reasonable bearing in mind that the bandwidth of the dynamics enhances for increasing the pump parameter, which in turn reduces the correlation

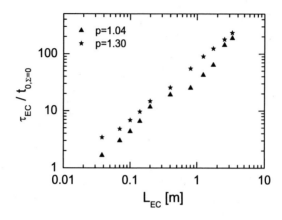

Fig. 3.13: Estimated dimension of the dynamics in dependence on L_{EC} and p.

properties of the dynamics. However, the estimated dimension merely gives an idea about the real dimension of the dynamics and its dependence on the relevant system parameters. Later on, we will present more precise measures that allow for estimation of the real dimension of the dynamics. Because of the technically limited acquisition bandwidth, we will apply these methods on numerically modeled intensity dynamics, obtained from a model that has been adapted to the experimental conditions. For the time being, we continue with the experimental characterization of the dynamics, but in the following, we focus on the spectral properties being associated with the intensity dynamics.

Spectral Properties

So far, we have focused on characterization of the intensity dynamics of ECSLs operating in the long cavity regime. However, our previous observations, illustrated in Figure 3.6, have shown that the emergence of irregular intensity dynamics is also in conjunction with the occurrence of linewidth broadening of the emitted light. This property is one reason why the corresponding regime of dynamics has been termed coherence collapse regime [118]. Although it seems intuitive to assume that the linewidth broadening inevitably implies a reduction of the coherence qualities of the SL, only sporadic efforts have been undertaken to characterize the specific coherence properties of ECSLs [136–138].

Before we explicitly analyze the coherence properties of ECSLs and its dependence on the system parameters, we briefly recall the main effects of delayed optical feedback (FB) on the spectral emission properties of SLs. These effects can be seen in Figures 3.4 d) and 3.5 b) in which we have shown optical spectra of a multimode and a singlemode SL for pronounced LFF intensity dynamics. The comparison of both spectra reveals two characteristic influences of FB on the spectral properties of the multimode SL. Firstly, we find that several longitudinal sidemodes are excited, while each of them exhibits a small redshift of $\Delta\lambda = 11$ pm with respect to the optical spectrum of the solitary laser. Secondly, the linewidth of the longitudinal modes of the SL is broader in the presence of optical feedback, when the SL exhibits LFF intensity dynamics. Figure 3.5 b) reveals similar results for the effect of FB on the spectral properties of a singlemode DFB SL which exhibits LFF intensity dynamics. In similarity to Figure 3.4 d), the corresponding optical spectrum of the DFB SL with FB also reveals a small redshift of the lasing LM of $\Delta\lambda = 87$ pm and linewidth broadening. Nevertheless, the redshift and the spectral broadening are more pronounced which is due to the fact that the DFB SL is subject to slightly stronger FB than the multimode SL. Despite the quantitative differences, the spectral properties of both types of SLs exhibit the same physical effects, i.e., feedback induced redshift and linewidth broadening. These two effects are representative for SLs and arise from the intrinsically strong nonlinearity, represented by the α-parameter.

Since the LMs of conventional multimode SLs are only weakly coupled and variations of the α-parameter within the affected spectral region can be neglected, all of the excited LMs exhibit similar redshift and linewidth broadening. In this case, the main difference between the spectral properties of singlemode and multimode SLs with optical feedback consists in the number of LMs that exhibit redshift and linewidth broadening. Therefore, multimode SLs provide potential for higher effective optical bandwidth, which suggests that for similar operation conditions the coherence properties of multimode SLs will reduce more than that of singlemode SLs. However, since the optical feedback induces intensity dynamics comprising various characteristic (slow and fast) time scales, the measured time-averaged optical bandwidth might not necessarily reflect the actual coherence properties. To answer this question, it is necessary to carefully study the influence of the intensity dynamics on the spectral properties of the SLs, and to directly measure the corresponding coherence properties. This task is particularly interesting with regard to multimode SLs with weakly coupled LMs, since for these SLs we can gain insight into the role of two characteristic spectral features for the coherence properties, i.e., the influence of the spectral dynamics in the vicinity of each of the LMs and the effects related to the occurrence of multimode emission. Thus, we benefit from this attractive offer and study the spectral properties of the HLP1400 multimode SL.

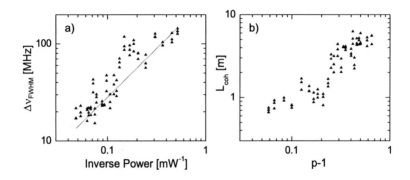

Fig. 3.14: Optical properties of the solitary HLP1400 semiconductor laser. Panel a) depicts
the linewidth of the SL in dependence on the emitted power. Panel b) presents
the corresponding coherence length, determined from the measured linewidth, as
a function of the reduced pump parameter $p - 1$.

In the following experiment, we measure the coherence length L_{coh} of the solitary
HLP1400 SL to provide a reference for the measurements with applied optical feedback. Therefore, we determine the linewidth of the SL for different pump parameters
with a Fabry-Perot scanning interferometer with a free spectral range of 2 GHz and a
finesse of $F > 200$. We find that for pump parameters of $p \geq 1.05$ the sidemodes of the
SL are mostly suppressed to better than 10 dB with respect to the dominant LM. For
theses conditions, we measure one single LM with the Fabry-Perot scanning interferometer. The lineshape of the dominant LM can be well approximated by a Lorentzian
from which we determine the linewidth $\Delta\nu_{FWHM}$, where FWHM stands for the full
width at half maximum. Figure 3.14 depicts the results for the spectral characterization of the solitary HLP1400 SL. In Figure 3.14 a), we present the measured linewidth
$\Delta\nu_{FWHM}$ as function of the inverse total output power of the SL. The double logarith-
mic representation reveals that the linewidth is inversely proportional to the emitted
power of the SL. This is highlighted by the gray line which we have fitted through the
data points. The fitting function corresponds to $\Delta\nu_{FWHM} = 265\,\text{MHz}\,\text{mW}/P$, with
P being the total output power of the SL measured in mW. This result is in good
agreement with the dependence of the spectral linewidth of a singlemode SL on its
output power, which is described by the modified Schawlow-Townes formula [55].

$$\Delta\nu_{FWHM} \;=\; \frac{h\nu v_g \alpha_m}{8\pi} \frac{R_{sp}}{P} (1 + \alpha^2) \qquad (3.12)$$

In this notation, ν represents the laser frequency, P is the output power of the SL, and v_g is the effective group velocity of the gain material. Furthermore, α_m denotes the mirror loss which is defined by $\alpha_m = (-1/L_{SL}) \ln R$, where R stands for the mirror facet reflectivity. R_{sp} is the spontaneous emission rate of the lasing mode including the astigmatism parameter. As it can be seen from Eq. 3.12, the linewidth of a solitary SL is directly related to the inverse of the total output power of the SL. The proportionality parameters depend on the specific properties of the SL. Amongst them, we find the α-parameter enhancing the linewidth as it is characteristic for SLs. We note, that for a multimode SL, which we have used in the experiment, the measured effective linewidth usually is also influenced by multimode effects [139]. However, in this experiment, the contribution of multimode effects is negligible, since the sidemode suppression of the solitary HLP1400 SL is higher than 15 dB for most of the accessible values of the pump parameter. Nevertheless, we also find a range of pump parameters for which the dynamics switches from one dominant LM to its neighboring LM. This is due to the wavelength dependence of the gain and the fact that the gain profile spectrally shifts for changing p. When two neighboring dominant modes are subject to similar gain they compete for common gain which effectively enhances the linewidths of the LMs [139]. This effect can be seen in Figure 3.14 a) for $P^{-1} \approx 0.2 \, \mathrm{mW}^{-1}$.

Now, that we have measured the linewidth of the solitary HLP1400 in dependence on the pump parameter, we are able to estimate the corresponding coherence length of the SL. For a Lorentzian spectral density, the coherence length L_{coh} can be determined from the following equation.

$$L_{coh} = \frac{c}{\pi \, \Delta\nu_{FWHM}} \qquad (3.13)$$

We utilize Eq. 3.13 to calculate L_{coh}. Figure 3.14 b) presents the results for the determined coherence length in dependence on the pump parameter, which has been reduced by the value for the solitary laser threshold. The figure discloses that L_{coh} grows linearly for increasing p. Furthermore, we estimate the accessible range of the coherence length to $0.75 \, \mathrm{m} < L_{coh} < 6.40 \, \mathrm{m}$. This range corresponds to common coherence lengths for multimode SLs without sophisticated resonator structure. Higher coherence lengths can be realized for DFB and DBR (Distributed Bragg Reflector) SLs for which the optimized resonator structure allows for coherence lengths of the order of 100 m. Short coherence lengths, on the other hand, can be realized by using anti-reflection coated multimode SLs. The coherence length of such SLs can be reduced to several millimeters. However, reduction of the facet reflectivity for decreasing L_{coh} is accompanied by reduction of the accessible output power.

Since optical feedback can induce linewidth broadening in SL systems, it provides an alternative technique for reducing the coherence length of SLs. In particular, this

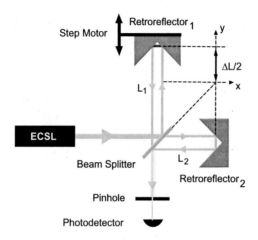

Fig. 3.15: Scheme of the Michelson interferometer setup for measuring the coherence length of the emitted light of ECSLs.

technique might also allow for realization of light sources with high output power, since the quality of the resonator is not reduced. We have already seen that the feedback induced linewidth enhancement of SLs is of dynamical origin in which various processes are involved covering a wide spectral and temporal range. Because of the underlying dynamics, the spectral emission properties of chaotically emitting ECSLs may considerably differ from that of solitary SLs, which usually exhibit a homogenously broadened spectral linewidth. Since, it can be expected that such influences on the linewidth are also accompanied by changes of the coherence properties, it is appropriate to directly measure the coherence properties of chaotically emitting ECSLs to obtain insight into the role of the underlying dynamics. This can be done by application of interferometric measurement techniques which allow for determination of the visibility function of the emitted light.

Figure 3.15 illustrates the Michelson interferometer setup which we use for characterization of the coherence properties of the HLP1400 SL with optical feedback. The light emitted by the ECSL hits a 50/50 beam splitter and is divided into two orthogonally propagating beams (1 and 2) with equal intensities. Each of the beams is reflected by a retroreflector and propagates back to the beam splitter. Subsequently, the beams interfere at the beam splitter with a path difference of $\Delta L = 2(L_1 - L_2)$. The path difference can be varied with a resolution of 100 nm within a range of $-12\,\mathrm{cm} < \Delta L < +12\,\mathrm{cm}$.

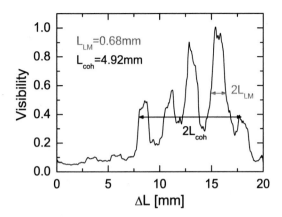

Fig. 3.16: Visibility function of the HLP1400 SL with delayed optical feedback and for the following operation conditions: $r = 0.063$, $p = 1.27$, and $\tau_{EC} = 2.9$ ns. The spacing of the peaks corresponds to twice the optical length of the SL cavity. Please note that in this measurement the zero-position of the path difference has been chosen arbitrarily.

In the experiment, we sweep the path difference from negative values to positive values with sub-wavelength resolution and monitor the intensity output of the interferometer. In this way, we obtain an interferogram that reflects the temporal coherence properties of the ECSL. Subsequently, we analyze the coherence properties in terms of the visibility function $VIS(\Delta L)$ which we calculate from the measured interferogram.

$$VIS(\Delta L) = \frac{P_{max}(\Delta L) - P_{min}(\Delta L)}{P_{max}(\Delta L) + P_{min}(\Delta L)} \tag{3.14}$$

Here, P denotes the power of the light measured at the output of the interferometer. $P_{max}(\Delta L)$ and $P_{min}(\Delta L)$ represent the nearest maximum and minimum of the interferogram at a path difference of ΔL. The maxima and minima of the interferogram correspond to constructive and destructive interference conditions between the electromagnetic fields of the two beams.

We utilize the visibility function for characterization of the spectral properties of the HLP1400 SL with optical feedback. Figure 3.16 depicts the experimentally obtained visibility function of the ECSL for LFF dynamics. In this case, the SL is subject to optical feedback with a delay of $\tau_{EC} = 2.9$ ns and with a feedback ratio of $r = 0.063$.

The pump parameter amounts to $p = 1.27$. The corresponding visibility function reveals several characteristics. Firstly, we identify peaks which are spaced by a path difference of $\Delta L_{peak} = 2.2\,\text{mm}$. This spacing is directly related to the spacing of the longitudinal modes of the SL, given by $\Delta\nu_{LM} = c/(2n_{eff}L_{SL})$, so that $\Delta L_{peak} = c/\Delta\nu_{SL}$. Secondly, the peaks reveal considerable but similar width which we denote by $2L_{LM}$. Thirdly, the envelope of the peaks describes a bell-shaped function which slightly reminds of a Gaussian distribution function.

This characteristic peaked visibility function significantly differs from typical visibility functions of continuously emitting thermal light sources or from the far reaching continuous visibility functions of solitary single mode lasers. To disclose the mechanisms leading to the observed characteristics, we study the influence of variations of the pump parameter and the feedback ratio on the visibility function. We find that these characteristics can be observed in a wide range of parameters within the coherence collapse regime of the intensity dynamics. Within this regime, the width of the peaks $2L_{LM}$ and the range of the envelope of the peaks depend on the operation parameters. Our results reveal the following dependencies:

- The width of the peaks slightly decreases for increasing the pump parameter and/or increasing the feedback ratio. The dependence of the width of the peaks is best pronounced for low pump parameters and weak feedback strength.

- The envelope of the peaks exhibits faster fall-off for increasing the pump parameter and/or increasing the feedback ratio. This dependence is also best pronounced for low pumping and weak feedback.

These results show characteristic influences of the operation parameters on the visibility function and therefore on the spectral properties. This sensitivity of the spectral properties is in agreement with our previous observation about the strong dependence of the intensity dynamics on the same operation parameters. Therefore, the results give evidence for a strong interrelation between intensity dynamics and spectral dynamics of ECSLs. By carefully analyzing this interrelation, we are able to associate spectral properties, manifesting themselves in the visibility function, with characteristics of the intensity dynamics.

From a comparative measurement of the visibility function of the solitary SL, we find that the occurrence of the pronounced peaks is related to the onset of noticeable multimode emission, which is induced by the optical feedback. For sufficiently strong optical feedback, the HLP1400 SL laser exhibits about six LMs with considerable modal power, as can be seen in Figure 3.4 d). The emission wavelengths of these modes, and also their spectral properties, are essentially determined by the resonance condition defined by the cavity of the SL. Therefore, the electromagnetic fields of the lasing LMs exhibit a fixed phase relation if measured at certain distances. In particular, high phase relation can be measured at distances that correspond to multiples of the optical length of the SL cavity, by which the LMs are defined. In particular, we find the same in-phase condition for the fields of all of the LMs for even multiples of the optical length of the SL, for $L = 2kL_{SL,opt}$, with $L_{SL,opt} = n_{eff}L_{SL}$ and k being an integer number. In contrast to this, we find anti-phase condition between the fields of adjacent LMs for $L = (2k - 1)L_{SL,opt}$. For that reason, the analysis of the coherence properties of the multimode ECSL discloses distinct peaks of high visibility at path differences corresponding to $L = 2kL_{SL,opt}$. However, since the linewidth of each of the LMs is broadened due to the feedback induced instabilities, the visibility of the peaks decreases for increasing path difference. In other words, the characteristics of the corresponding envelope of the peaks is related to the linewidth of the LMs. As a consequence thereof, the envelope exhibits faster fall-off for enhanced linewidth of the LMs which can be realized by increasing the feedback strength and by choosing moderate pump parameters so that the ECSL operates well within the coherence collapse regime. The width of the peaks of the visibility function and the minimal visibility between the peaks, on the other hand, depends on two properties. Firstly, the visibility between the peaks substantially decreases for increasing the number of lasing LMs. This effect can be explained in terms of multiple beam interference for which the coherent superposition of many fields with different wavelengths gives rise to low visibility. This is due to the fact that the average phase of the fields added up to a certain point cancels out. Secondly, we find also an influence of the dynamics on the width of the peaks. For increasing feedback strength, the spectral range of the dynamics enhances according to Eq. 3.9. Therefore, the linewidth of each of the LMs increases which results in a faster decrease of the visibility at the peaks at $L = 2kL_{SL,opt}$. In similarity to this effect, the width of the visibility peaks also decreases for increasing the pump parameter when the bandwidth of the intensity dynamics enhances. Since the intensity dynamics is correlated with the spectral dynamics, the bandwidth of the spectral dynamics also enhances reducing the visibility.

The results show that the combined occurrence of multimode emission and pronounced spectral dynamics drastically influences the visibility properties of the light emitted by the SL. These two effects cause the characteristic shape of the visibility function, which is typical for chaotically emitting multimode ECSLs. Furthermore, we have revealed that the visibility properties can be influenced by variation of the relevant system parameters, i.e., the feedback strength and the pump parameter. This is an interesting result, since it offers potential for tuning of the coherence length of the SL system via adjusting its nonlinear dynamics. To verify this idea, we need to measure the coherence length of the ECSLs in dependence on the relevant system parameters. Conventionally,

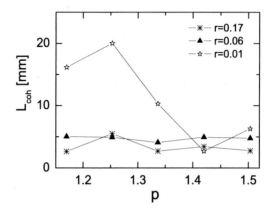

Fig. 3.17: Coherence length for the HLP1400 semiconductor laser in dependence on the feedback ratio r and the injection parameter p. The external cavity length amounts
to $\tau_{EC} = 2.9\,$ns. Please not that the lines only serve as guides to the eye.

the coherence length is defined by the maximum interferometric path difference for
which the visibility equals $1/e$. We adopt this definition for determination of an upper
found of the coherence length L_{coh}. Therefore, we interferometrically measure the
visibility function of the emitted light for different pump parameters and feedback
rates. For each of the visibility functions, we determine the maximal range $2L_{coh}$ for
which the visibility is higher than $1/e$, as it is illustrated in Figure 3.16. Half of the
width of this range corresponds to the coherence length L_{coh} of the emitted light.

In Figure 3.17, we present the determined coherence length of the HLP1400 with optical
feedback in dependence on the pump parameter. In the experiment, the delay time
corresponds to $\tau_{EC} = 2.9\,$ns. The figure depicts results for three different values of
the feedback ratio of $r = 0.17$, $r = 0.06$ and $r = 0.01$, covering a wide range of the
coherence collapse dynamics regime. For the accessible dynamics in this regime, we
find coherence lengths in the range of between $3\,$mm $< L_{coh} < 20\,$mm. This range of
L_{coh} discloses a reduction of the coherence length of up to three orders of magnitude
with respect to the corresponding coherence length of the solitary SL, which we have
determined to be $0.75\,$m $< L_{coh} < 6.40\,$m. Furthermore, the accessible range of the
coherence length gives rise to the possibility to tune the coherence length within one
order of magnitude by controlled manipulation of the dynamically relevant system
parameters. In agreement to our previous findings for the dependencies of the visibility
function on the system parameters, we find that the coherence length decreases for

increasing the pump parameter, when the dynamics evolves into the fully developed coherence collapse regime. This can best be seen for weak feedback of $r = 0.01$. For low pumping of $p = 1.25$, we find comparably long coherence length of about $L_{coh} \approx 20$ mm for intensity dynamics that exhibits small dynamical bandwidth in the rf spectra. For continuously increasing p to $p = 1.5$, the coherence length distinctly reduces to $L_{coh} = 6$ mm, while the bandwidth of the rf spectra of the dynamics increases. The dynamics for higher feedback ratios of $r = 0.06$ and $r = 0.17$ reveals broadband rf spectra indicating that the ECSL operates well within the coherence collapse regime, even for low pump parameters. For this dynamics, we measure comparably short coherence lengths of the order of $L_{coh} \approx 5$ mm, while L_{coh} only slightly deceases, if we increase the pump parameter from $p = 1.2$ to $p = 1.5$. In agreement with the observed small dependence of L_{coh} on p, the corresponding spectral properties reveal only a moderate increase of the bandwidth of the intensity dynamics. However, it is worth noting that the comparison of the results for $r = 0.06$ and $r = 0.17$ demonstrates shorter coherence length for stronger feedback.

The analysis of the coherence properties has revealed two essential mechanisms which determine the coherence properties of chaotically ECSLs. Firstly, the coherence length sensitively depends on the feedback strength. This is due to the fact that the feedback strength determines the optical spectral range of the dynamics, which we know from the analysis of the Lang-Kobayashi SL rate-equations. Therefore, the optical bandwidth enhances for increasing the feedback strength. This results in broadening of the linewidth and, hence, it reduces the coherence length. Secondly, the coherence length also depends on the pump parameter which influences the bandwidth of the intensity dynamics. However, this dependence is only well pronounced when the dynamical properties considerably change for variation of p. This is the case for $r = 0.01$, for which the dynamics evolves from weakly pronounced spectrally confined dynamics to well pronounced spectrally broadband dynamics. In contrast to this, we find only small influence of p on L_{coh} for dynamics well within the coherence collapse regime. The origin of this behavior can also be explained in terms of spectral dynamics. The characteristic linewidth broadening in the coherence collapse regime is due to the fact that the dynamics spans most of the accessible optical bandwidth, which is defined by the feedback strength. The corresponding mechanism that influences the coherence length for variation of p can only be of temporal nature, since the pump parameter does not influence the spectral extension of the attractor, as can be seen from Eq. 3.6. Indeed, we find that the coherence length reduces for enhancing the spectral bandwidth of the intensity dynamics which is realized by increasing the pump parameter. Hence, the dynamics comprises faster frequency components and evolve more irregularly on the attractor which is also extended in frequency space. This increasing irregularity of the spectral dynamics, in turn, can explain the observed reduction of the coherence length.

Therefore, the dependence of L_{coh} on p is less pronounced for ECSLs operating well within the fully developed coherence collapse regime. In this case, the enhancement of the linewidth by application of stronger feedback more efficiently reduces the coherence length. Nevertheless, the results we have presented in Figure 3.7 demonstrate that it is not possible to reduce L_{coh} to any desired value by increasing r, since the dynamics stabilizes for high feedback strength via locking to the external resonator.

In summary, the presented characterization of the spectral properties of chaotically emitting ECSLs which operate in the long cavity regime have revealed that the spectral dynamics is connected to the intensity dynamics. Our results reveal that the spectral bandwidth of the dynamics enhances for increasing the feedback strength. In particular, we have analyzed the characteristic coherence properties of chaotically emitting ECSLs. The results demonstrate that the coherence length of SLs can be reduced by more than three orders of magnitude for dynamics in the coherence collapse regime. This property opens up potential for the controlled manipulation of the coherence properties of SLs by utilization of their nonlinear dynamical properties. This possibility is very attractive for applications which require light sources with well-defined, or even tunable, coherence properties. We will address this very interesting topic in Chapter 4, in which we investigate the possibility of harnessing the NLD of SLs for tailoring the emission properties of SL systems.

The experimentally observed interrelation of the optical properties and the intensity dynamics represents an interesting result. This is not only due to the fact that it allows for predictions about the optical properties of ECSLs by monitoring the intensity dynamics, but even more it reflects characteristic properties of the nonlinear dynamics of chaotically emitting ECSL systems. The experiments did not only demonstrate that the dynamics comprises faster frequency components for higher pump parameters, but even more the results suggest that the dynamics also becomes more irregular. This finding reveals that further insight into the properties of the nonlinear dynamics of ECSLs is required to understand the emergence and development of dynamics in dependence on the system parameters. Accordingly, it is desirable to characterize the dynamics in terms of NLD measures, such as Lyapunov exponents, information dimension, and entropy. Unfortunately, at the present time, such characterization cannot be accomplished experimentally, because of existing limitations in detection bandwidth. However, with the Lang-Kobayashi SL rate-equations we have an excellent model at hand that provides a beneficial alternative to gain the desired insight into the dynamical properties of chaotically emitting ECSLs.

Modeling Results

In this subsection, we present results obtained for numerical modeling of the dynamics of ECSLs using the Lang-Kobayashi SL rate-equations. Numerical modeling can provide beneficial complementary insight into the dynamical properties of ECSLs, since it allows for computation of time series of the dynamics with appropriate temporal resolution. In this way, it is possible to consider the full bandwidth of dynamics, which is presently beyond experimental possibilities. Premise is that the model is able to efficiently reproduce the relevant dynamical effects of the experiment. In this context, the Lang-Kobayashi model has proven its suitability for modeling the dynamics of conventional ECSLs. Therefore, the Lang-Kobayashi model can provide further insight into the statistical properties of the dynamics of ECSLs by application of analysis methods of NLD onto the modeled time series.

In the following, we briefly introduce the NLD measures which we use for analysis of the statistical properties of the dynamics of ECSLs. In particular, we present results for the Lyapunov exponents, denoted as λ_i, and the Kaplan-Yorke dimension of the dynamics, which we refer to as d_{KY}. Later on in this chapter, we will also study the entropy of the dynamics for which we calculate the Kolmogorov-Sinai entropy, termed as h_{KS}. The presented numerical results have been obtained within the collaboration between our group in Darmstadt, and the UIB as well as IMEDEA, Palma de Mallorca, Spain. The numerical modeling has been performed by R. Vicente et al. [140].

The Lyapunov exponents (λ_i) of some dynamics provide a useful measure for the rate at which two initially nearby states of the dynamics diverge as the dynamics evolves. The Lyapunov exponents of delay differential equations can be calculated from the time series of the dynamics according to the method proposed by Farmer [141]. Therefore, the delay differential equations, which in our case are the Lang-Kobayashi (LK) SL rate-equations, need to be numerically integrated. Technically, this is done by application of an Adam-Bashforth-Moulton fourth-order predictor-corrector method. Since the LK equations are delay differential equations which are mathematically infinite dimensional, the dynamics potentially exhibits an infinite number of Lyapunov exponents. However, numerically it is only possible to determine the largest Lyapunov exponents of the infinite number of exponents. Nevertheless, the largest Lyapunov exponents represent the most relevant exponents describing the statistical properties of the dynamics. In particular, we are interested in the positive Lyapunov exponents of the dynamics since they reveal diverging dynamics, or equivalently, they indicate chaotic dynamics. Furthermore, the number of positive Lyapunov exponents can be treated as a rough estimate for the dimension of the strange attractor. In this way, the Lyapunov spectrum can provide fundamental insight into the dynamics allowing

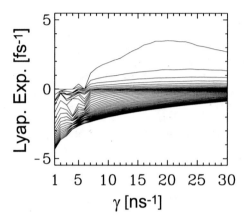

Fig. 3.18: Largest Lyapunov exponents as a function of the feedback rate for $\tau_{EC} = 1\,\text{ns}$ and $p = 1.5$. Courtesy of R. Vicente.

for characterization of geometrical and dynamical aspects of the corresponding strange attractor [142].

We benefit from this complementary insight into the dynamics of ECSLs operating in the long cavity regime and analyze the Lyapunov exponents of the dynamics in dependence on the system parameters. Therefore, we numerically integrate the Lang-Kobayashi SL rate-equations as they are presented in Eqs. 3.3, 3.4, and 3.5. For now, we do not consider the Gaussian white noise term $\sqrt{2\beta\gamma_N N(t)}$ of Eq. 3.4, since we focus on the properties of the deterministic dynamics. We use a set of parameters for the modeling that reflect the properties of conventional single mode SLs, such as the IPAG DFB SL. The linewidth enhancement factor amounts to $\alpha = 5$, the differential gain is $\xi = 1.5 \times 10^{-8}\,\text{ps}^{-1}$, the cavity decay rate is $\Gamma = 500\,\text{ns}^{-1}$, the carrier life time corresponds to $T_N = 2\,\text{ns}$, the transparency carrier number is $N_0 = 1.5 \times 10^8$, and the gain suppression coefficient is $\epsilon = 5 \times 10^{-7}$. Additionally, we consider the delay time to be $\tau_{EC} = 1\,\text{ns}$ and the pump parameter being $p = 1.5$. We integrate the LK equations for this set of parameters for increasing the feedback rate from $\gamma = 0.1\,\text{ns}^{-1}$ to $\gamma = 30\,\text{ns}^{-1}$. This range of feedback rate corresponds to ratios of approximately $0 \leq r < 0.08$. Then, we calculate the Lyapunov exponents for each of the time series.

Figure 3.18 presents the spectrum of the largest Lyapunov exponents which we obtain for these conditions. For low feedback rates of up to $\gamma = 4\,\text{ns}^{-1}$, we do not find positive Lyapunov exponents since the emission is continuous for very weak feedback and periodic at the onset of dynamics. We identify one zero-valued Lyapunov exponent persisting for the whole range of feedback ratios, which implies neutral stability. This zero Lyapunov exponent is associated with the continuous symmetry of the system under shift of the optical feedback phase. At about $\gamma \approx 5\,\mathrm{ns}^{-1}$, we find that the largest Lyapunov exponent becomes positive. This transition demonstrates that the dynamics of ECSLs operating in the coherence collapse regime indeed is of chaotic nature. Furthermore, the spectrum discloses a continuously increasing number of positive Lyapunov exponents for increasing feedback ratio. This property already suggests that the dimension of the dynamics grows for increasing the feedback strength, since each of the Lyapunov exponents represents the divergence in one orthogonal direction of the phase space of the dynamics. Interestingly, we find that the positive Lyapunov exponents exhibit a maximum value for moderate feedback rates. This maximum is particularly pronounced for the largest Lyapunov exponent for which we identify a maximum at $\gamma \approx 20\,\mathrm{ns}^{-1}$. Hence, the exponential divergence of nearby states of the dynamics is largest in the corresponding direction of phase space for this feedback strength. For feedback ratios beyond this extremum, the value of the positive Lyapunov exponents slowly decreases so that further increasing the feedback strength does not give rise to larger values of the Lyapunov exponents. This property has significant consequences for the predictability of the dynamics, which we will discuss later.

Similar to the experimental characterization of the dynamics in the coherence collapse regime, we also numerically study the influence of the pump parameter p on the dynamical properties. Therefore, we set the feedback rate to $\gamma = 20\,\mathrm{ns}^{-1}$ and integrate the LK equations for different pump parameters in the range of between $1 < p < 3$. Then, we calculate the Lyapunov spectra from the obtained time series. Figure 3.19 illustrates the dependence of the largest Lyapunov exponents on the pump parameter. The figure reveals that the largest Lyapunov exponent initially increases for incrementing the pump parameter from $p = 1$ to $p = 1.5$, where it reaches its maximum value. For further incrementing the pump parameter, the largest Lyapunov exponent monotonically decreases. The second largest Lyapunov exponent exhibits similar dependence on the pump parameter, but with a maximum that is slightly shifted toward higher pump parameters. Similar behavior can be observed for the consecutively smaller positive Lyapunov exponents, which also reveal less dependence of their magnitude on the pump parameter. As a consequence thereof, the results give evidence for the existence of a maximal degree of irregularity of the dynamics which can be achieved by adjusting the pump parameter.

Nevertheless, to allow for concrete statements about the properties of chaotic dynamics it is essential to provide a quantitative measure for the degree of disorder. Such a measure is defined by the information dimension. The information dimension describes

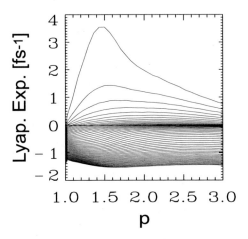

Fig. 3.19: The 50 largest Lyapunov exponents as a function of the pump parameter. The external round trip time is $\tau_{EC} = 1\,\text{ns}$ and the feedback rate amounts to $\gamma = 20\,\text{ns}^{-1}$. Courtesy of R. Vicente.

how the amount of information which is needed to locate the state of the system with defined precision scales with the required accuracy. Unfortunately, the computational effort for determination of the information dimension of high-dimensional delay dynamics of the LK equations is beyond today's computational capabilities. Therefore, it is convenient that according to the Kaplan-Yorke conjecture the information dimension can be estimated by the Kaplan-Yorke dimension, denote as d_{KY}, which can be numerically computed from the Lyapunov spectrum:

$$d_{KY} = j + \frac{\sum_{i=1}^{j} \lambda_i}{|\lambda_{j+1}|} \tag{3.15}$$

Here, the Lyapunov exponents have been ordered considering their magnitude, $\lambda_i \geq \lambda_{i+1}$. The integer j expresses the number of degrees of freedom of the dynamics, given by the constraint: $\sum_{i=1}^{j} \lambda_i \geq 0$ and $\sum_{i=1}^{j+1} \lambda_i < 0$.

Equation 3.15 allows us to quantitatively estimate the geometric dimension of the chaotic attractor from the Lyapunov exponents of its dynamics[2]. The procedure is

---

[2]Please note that the absolute value of the estimated dimension of a high-dimensional chaotic attractor can depend on the numerical method which is applied for its determination.

straightforward since we have already calculated the Lyapunov spectra for various operation conditions of the ECSL. Figure 3.20 provides an overview over the dependence
of the Kaplan-Yorke dimension (d_{KY}) on the relevant system parameters of the ECSL.
In Figure 3.20 a), we present the influence of the feedback rate on the Kaplan-Yorke
dimension for two different delay times τ_{EC}. The data marked by the diamonds corresponds to dynamics in the long cavity regime for which the external cavity round
trip time amounts to $\tau_{EC} = 1\,$ns. Additionally, we have plotted the computed Kaplan-
Yorke dimension for a short cavity for which $\tau_{EC} = 0.2\,$ns, which is marked by the
crosses. The results disclose that for both cases d_{KY} almost linearly grows as soon as
the feedback rate exceeds a value of about $\gamma = 7\,$ns^{-1}. The Kaplan-Yorke dimension for
the long cavity regime dynamics reveals high-dimensionality which verifies that the dynamics in the long-cavity regime is indeed hyper-chaotic. In the case of the short cavity,
we find that the attractor of the dynamics exhibits less high-dimensionality. However,
even for the short delay time of $\tau_{EC} = 200\,$ps, d_{KY} exhibits attractor dimensions of
between $4 < d_{KY} < 9$. Hence, short ECSLs can also exhibit hyper-chaotic dynamics
for operation parameters corresponding to operation in the coherence collapse regime.
Figure 3.20 b) illustrates the dependence of d_{KY} on the pump parameter p for a range
of $1 \leq p \leq 3$. The data reveals that the attractor dimension grows steadily when we
start to increase p from $p = 1$ until d_{KY} reaches a maximum at around $p \approx 2.2$, while
d_{KY} starts to slowly decrease for further increasing p. Finally, in Figure 3.20 c), we
present the influence of the delay time on the magnitude of d_{KY}.[2] In similarity to the
Figures 3.20 a) and b), the figure reveals that d_{KY} stays at zero level until a certain
threshold value is reached. The dynamics suddenly becomes chaotic for $\tau_{EC} \approx 150\,$ps
which is indicated by the jump of the Kaplan-Yorke dimension to $d_{KY} \approx 4$. For longer
delay times, we find a linear dependence of the Kaplan-Yorke dimension on the delay
τ_{EC}. Therefore, it is possible that the Kaplan-York dimension of ECSLs with long
external cavities can easily exceed $d_{KY} > 100$, as it has been reported by Ahlers et
al. [134]. The observed linear dependence of d_{KY} on the delay time is in qualitative
agreement with the experimentally estimated dimension of the dynamics, which we
have calculated from the measured auto-correlation properties of the intensity dynamics, according to the Le Berre conjecture. The comparison of the numerically obtained
dimension, depicted in Figure 3.20 c), and the experimentally estimated dimension,
shown in Figure 3.13, reveals high-dimensional dynamics for both measures. However,
a close look at the results reveals slightly lower dimension for the experimental results.
One reason for this can be found in the different pump parameters for the experiment
and the modeling. The experimental results demonstrate that the estimated dimension
for $p = 1.30$ is higher than that for $p = 1.08$. Again, this result is in good agreement
with the numerically revealed growing of d_{KY} for increasing pump parameter. Since
the pump parameter in the modeling amounts to $p = 1.5$, while the pump parameter

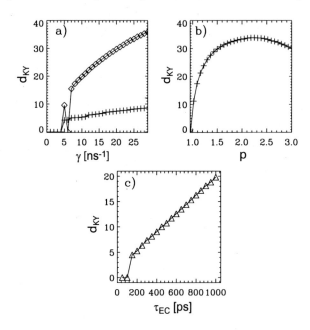

Fig. 3.20: Panel a) presents the Kaplan-Yorke dimension as a function of the feedback rate
for $\tau_{EC} = 1\,\text{ns}$ (diamonds) and $\tau_{EC} = 0.2\,\text{ns}$ (crosses), and for $p = 1.5$. The
Kaplan-Yorke dimension as a function of the pump parameter is illustrated in
panel b), where the external round trip time is $\tau_{EC} = 1\,\text{ns}$ and $\gamma = 20\,\text{ns}^{-1}$.
Panel c) depicts the influence of the delay on the Kaplan-Yorke dimension. The
feedback rate has been set to $\gamma = 10\,\text{ns}^{-1}$, while the pump parameter corresponds
to $p = 1.5$. Courtesy of R. Vicente.

in the experiment was only $p = 1.30$, one could also expect slightly higher dimension
for the experimental estimates for $p = 1.5$. Unfortunately, the limited bandwidth of
the oscilloscope did not allow for reliable experimental determination of the dimension of the dynamics for higher pump parameters. Nevertheless, the comparison of
the experimental and the numerical results clearly demonstrates that meanwhile it is
experimentally possible to gain fundamental insight into the dynamical properties of
high-dimensional chaotic dynamics.

The numerically obtained results highlight that modeling represents a fruitful complementary approach for analysis of the dynamical properties of chaotically emitting

ECSLs. In particular, we have verified that the dynamics of ECSLs operating in the coherence collapse regime indeed is chaotic, and even more we have revealed its high-dimensionality. Our results have disclosed interesting dependencies of the magnitude of the Lyapunov exponents and the Kaplan-Yorke dimension on the relevant system parameters. Firstly, the magnitude of the largest Lyapunov exponent can be maximized for moderate pumping and moderate feedback conditions, while further increasing of both parameters degrades its magnitude. Secondly, we find that the Kaplan-Yorke dimension of the dynamics can also be maximized for moderate pump parameters, while we do not find a maximum for d_{KY} for varying the feedback rate. In contrast to this, the dimension continuously grows for increasing the feedback strength. Likewise, the dimension linearly grows for increasing delay time. These results imply that the complexity of the dynamics distinctly varies within the coherence collapse regime. Even more, the results suggest the existence of a set of parameters for which the dynamics exhibits maximal complexity. In this context, we point out that, pump parameters of about $p = 1.5$ and moderate feedback strength in the range between $0.1 < r < 0.2$ are basis for high maximum positive Lyapunov exponents. We also note that at this point predictions about an optimal set of parameters are rather speculative, since further knowledge about the dynamical properties and dependencies is required.

So far, we have concentrated on the analysis of the dynamical properties of ECSLs operating in the long cavity regime. Nevertheless, we have just seen that ECSLs with short cavities also give rise to high-dimensional chaotic dynamics, if they are operated under otherwise comparable conditions so that the control parameters correspond to that of the coherence collapse regime, i.e., moderate feedback strength and sufficient high pump parameters. For ECSLs with sufficiently short cavities, the external cavity round trip frequency ν_{EC} can be of the same order than the relaxation oscillation frequency ν_{RO}, or even higher. Since our results demonstrate that both frequencies play a key role for the dynamics of ECSLs, this reversal in the hierarchy of relevant frequencies implicates the occurrence of physically different dynamical phenomena. We address this interesting question in the subsequent chapter in which we study the dynamical properties of chaotically emitting short ECSLs.

3.4 Chaotic Dynamics of Lang-Kobayashi Systems: The Short Cavity Regime

Recently, the dynamics of SLs with short external cavities has attracted much interest. The term "Short Cavity" implies that the round trip time of the external cavity τ_{EC} is of the order of or shorter than the period of the relaxation oscillations. This reversal in the hierarchy of relevant time scales gives rise to different dynamical behavior of short ECSL systems, if compared to that of conventional long cavity regime ECSL systems. In particular, it has been demonstrated that the intensity dynamics in the Short Cavity Regime (SCR) can exhibit fundamentally interesting dynamical phenomena comprising regular and irregular Pulse Packages (PP) which have not been observed in the Long Cavity Regime (LCR). Furthermore, it has turned out that the dynamics in the SCR sensitively depends on the feedback phase $\Delta\Phi$. However, it was just five years ago when Heil et al. presented the first systematic experimental characterization of the dynamics of ECSLs operating in the SCR [86]. The authors revealed a characteristic scenario comprising stable emission, periodic, quasi-periodic and chaotic states for varying the phase of the feedback for low levels of pumping. Subsequent bifurcation analysis of the SCR dynamics highlighted fundamental differences of SCR dynamics if compared to that of LCR [87, 143].

The investigations in the SCR have so far mainly focused on the emergence and the development of the dynamics which can be observed for low levels of pumping, near the solitary laser threshold current. The dynamical phenomena which have been studied in this regime comprised regular, quasi-periodic or low dimensional chaotic dynamics. Nevertheless, it has also been reported that SCR dynamics can also exhibit pronounced irregular behavior for moderate pump parameters [86]. However, up to now, detailed studies of the dynamical properties in this regime are still lacking. This is mainly due to the fact that the dynamics in the SCR can comprise frequencies which are well beyond the relaxation oscillation frequency ν_{RO}. Therefore, the bandwidth of the dynamics can easily reach up to several $10\,\text{GHz}$ [86, 87] which previously hampered experimental characterization of the dynamics in this regime.

In the following, we study the dynamical properties of chaotically emitting ECSLs operating in the SCR. We verify the possibility to generate high-dimensional broadband chaos within the SCR. With this aim, we experimentally study the influence of the pump parameter p, the feedback ratio r, the delay time τ_{EC} and, in particular, the feedback phase $\Delta\Phi$ on the emission dynamics. We apply complementary experimental and numerical analysis methods to gain insight into the dynamical properties of well-pronounced chaotic dynamics generated by short ECSL systems.

3.4.1   From Regular to Irregular Pulse Packages

In this subsection, we experimentally study the properties of the intensity dynamics of ECSLs operating in the SCR. First, we introduce the characteristic pulse package dynamics phenomenon which can be observed in the short cavity regime. On this basis, we discuss particularities of the dynamics in the SCR and show how it is possible to achieve broadband chaotic intensity dynamics, which will be subject to our subsequent analysis.

We use the same experimental setup as for the characterization of the dynamics in the LCR, which we have described in Subsection 3.1.1. The only difference is that we drastically reduce the length of the external cavity down to the range of centimeters. Again, we utilize the HLP1400 SL for most of the experiments. However, we have also performed complementary experiments with single mode SL (IPAG and Optospeed) for which we find qualitatively agreeing results, similar to the LCR dynamics. In contrast to the experiments in the LCR, in the SCR special care has to be taken for the stability of the system, since the dynamics sensitively depends on the optical feedback phase $\Delta\Phi$ [86,87]. This is in contrast to the behavior of the dynamics in the LCR, where the dynamics is almost independent on small variation of $\Delta\Phi$. In the LCR, an influence of a change of $\Delta\Phi$ on the dynamics can only be revealed by application of detection techniques which utilize synchronization phenomena [144].

In the experiment, we set the length of the external cavity to $L_{EC} = 3.9$ cm, corresponding to a round trip frequency of $\nu_{EC} = 3.8$ GHz. This length has been chosen to allow for resolution of the external cavity frequency with the oscilloscope which has an analog bandwidth of 4 GHz. We have applied moderate feedback with a ratio of $r = 0.048$ for which we measure a threshold reduction of $1 - I_{th,fb}/I_{th,sol} = 0.053$. We operate the laser near the solitary laser threshold current at $p = 1.08$. For this pump level, the relaxation oscillation frequency of the solitary laser has been determined to be $\nu_{RO} \approx 1.8$ GHz, which is substantially smaller than the external cavity round trip frequency being $\nu_{EC} = 3.8$ GHz. Thus, the condition for operation within the SCR is satisfied $\nu_{RO} < \nu_{EC}$. Consequently, we find the characteristic sensitivity of the dynamics on the feedback phase. For increasing $\Delta\Phi$, we observe the cyclic dynamics scenario as it has been described by Heil et al. [86,87] in which the dynamics evolves from stable emission via *Hopf Bifurcation* to periodic states, followed by a *Torus Bifurcation* leading to quasi-periodic states, from which the dynamics may develop in a quasi-periodic route to weakly pronounced chaotic states until the attractor collapses in a *Boundary Crisis*, and the state of stable emission is reached again.

Since we are interested in the dynamical behavior of the ECSL, we adjust to the most pronounced dynamics we can achieve for these operation conditions. Then, the intensity dynamics exhibits regular pulse packages. This characteristic dynamical state of the SCR is depicted in Figure 3.21 a). The intensity dynamics disclose fast regularly spaced pulses revealing two relevant time scales. Firstly, the fast pulsations are separated by approximately $\Delta t = 260$ ps which corresponds to a repetition frequency of $1/(\Delta t) = 3.8$ GHz. This frequency agrees with the external cavity round trip frequency being $\nu_{EC} = 3.8$ GHz. This means that the dynamics of the fast pulsations in the SCR

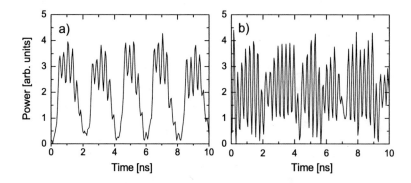

Fig. 3.21: Transition from regular pulse package dynamics, presented in panel a), to irregular
pulse package dynamics, shown in panel b), for increasing the pump parameter in
the short cavity regime. In panel a) $p = 1.08$, while $p = 1.30$ in panel b). Other
conditions are: $L_{EC} = 3.9\,\mathrm{cm}$, $\tau_{EC} = 260\,\mathrm{ps}$, and $r = 0.05$.

is dominated by the delay, which is in contrast to the LCR for which the fast pulsations
are mainly dominated by the relaxation oscillation frequency. The second relevant time
scale which we can identify from the time series is defined by the period of the regular
pulse package (RPP). In this case, we can estimate a period of the PP of $T_{RPP} \approx 2\,\mathrm{ns}$
which corresponds to $\nu_{RPP} \approx 500\,\mathrm{MHz}$. This time scale is much slower than that of
the delay time. We find that the RPP frequency increases linearly for increasing the
pump parameter up to $p = 1.25$, while it decreases with an inverse fourth order power
law dependence for increasing the feedback ratio up to $r = 0.15$, as it has also been
reported by Heil et al. [49]. However, the qualitative shape of the envelope remains the
same for changes within this range of parameters. The frequency of the fast pulsations
remains essentially constant for changing the feedback strength and the pump current.

In contrast to this, the dynamical behavior can drastically change when we further
increase the pump parameter to $p = 1.30$. For this pump parameter, the relaxation
oscillation frequency of the solitary laser corresponds to $\nu_{RO} \approx 2.8\,\mathrm{GHz}$, so that the
ECSL still operates well within the SCR. We find that the dynamics for moderate
pump parameters also sensitively depends on $\Delta\Phi$, but now the dynamics gives rise to
an abrupt transition from stable emission to chaotic dynamics until it eventually jumps
back to stable emission. In Figure 3.21, we present pronounced chaotic intensity which
we achieve for carefully adjusted feedback phase. The dynamics reveal irregular fast
pulsations on the time scale of the delay, so that we cannot identify pulse packages
anymore. A closer look at the time series reveals that not only the amplitude of the

pulses considerably varies from pulse to pulse, but also the interval between the pulses jitters slightly. This demonstrates that it is possible to destroy the regularity of the dynamics by increasing the pump parameter. Consequently, the SCR can also exhibit very irregular dynamics for moderate feedback and sufficiently high pump parameters, which is similar to our results for the fully developed coherence collapse regime in the LCR.

3.4.2 Properties of Chaotic Short Cavity Regime Dynamics

In this subsection, we present the first comprehensive characterization of the dynamical properties of irregular SCR regime dynamics. We analyze the dynamical properties in dependence on the relevant system parameters using different experimental methods. Additionally, we present results obtained from numerical modeling which provide meaningful insight into the nonlinear dynamics of chaotically emitting short ECSLs. Our analysis discloses fundamental differences between chaotic SCR dynamics and chaotic LCR dynamics.

Bandwidth of Dynamics

The time series in Figure 3.21 demonstrated that it is possible to achieve irregular intensity dynamics in the SCR by increasing the pump parameter and readjusting the feedback phase for dynamics. Nevertheless, the presented short segments of the intensity dynamics only provide limited insight into the properties of the dynamics, since relevant information about the long-time behavior of the dynamics is not accessible. To circumvent this, we need to study the statistical properties of the dynamics, which is of particular interest when the dynamics comprises a wide range of frequencies. Therefore, we measure the rf spectra of the intensity dynamics in the SCR from which we can identify characteristic system frequencies providing insight into relevant mechanisms of the dynamical properties.

Figure 3.22 b) depicts the rf spectra of the intensity time series of Figure 3.21. The spectrum for the regular pulse package dynamics near the solitary laser threshold for $p = 1.08$ is presented in Figure 3.22 a). We find several peaks, which we can assign to the two characteristic frequencies, we have identified from the time series. On the one hand, we can identify a sharp peak at the external cavity round trip frequency being $\nu_{EC} = 3.8\,\mathrm{GHz}$. This represents the dominant frequency of the fast intensity pulsations. On the other hand, we can also identify the characteristic frequency of the envelope of the regular pulse package, which we find at $\nu_{RPP} \approx 500\,\mathrm{MHz}$. The corresponding peak is almost as sharp as that of the fast intensity pulsations underlining the regularity of

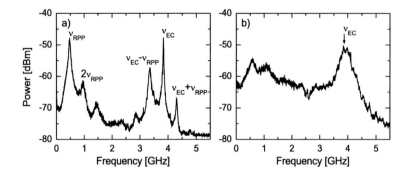

Fig. 3.22: rf spectra corresponding to the intensity dynamics depicted in Figure 3.21. Panel a) reveals quasi-periodic dynamics being characteristic for regular pulse package dynamics observed for low pump parameters of $p = 1.08$. The rf spectrum for $p = 1.30$ is depicted in panel b) revealing that the bandwidth enhances for increasing p. The other parameters are given in the captions of Figure 3.21.

the pulse package. Furthermore, we identify harmonics of ν_{RPP} and sum and difference frequencies of ν_{RPP} and ν_{EC}. Hence, the rf spectrum exhibits typical characteristics for quasi-periodic dynamics. From Chapter 2, we remember that the quasi-periodic route is one of the classical route leading to chaos. Indeed, the width of the peaks in the rf spectrum may already suggest the presence of weakly pronounced chaotic dynamics. Even more, this observation also indicates that more broadband chaotic dynamics might be achievable for tuning the right system (bifurcation) parameter. Finally, we point out that the spectrum reveals dynamics well beyond the relaxation oscillation frequencies of the solitary SL being $\nu_{RO} = 1.8\,\text{GHz}$. The rf spectrum does not give rise to dynamics components at that frequency. However, we note that the relaxation oscillation frequency of the feedback system can be shifted considerably to higher frequencies. Nevertheless, this not obvious presence of relaxation oscillations is in clear contrast to our findings for the coherence collapse dynamics in the LCR, for which the relaxation oscillation frequency represented one of the dominant system frequencies determining the properties of the dynamics.

Next, we investigate the influence of the pump parameter on the spectral properties of the intensity dynamics. Figure 3.22 b) shows the rf spectrum for the irregular intensity dynamics which we have been able to achieve for sufficiently high pump parameters of $p = 1.30$. The figure reveals a continuous, almost flat rf spectrum comprising high dynamical bandwidth with maximum frequencies of up to $5\,\text{GHz}$. These interesting spectral properties drastically differ from that of the quasi-periodic dynamics we have identified for $p = 1.08$. In contrast to the regular dynamics for low levels of pumping, the dynamics for sufficiently high pump parameters does not give rise to apparent characteristic frequencies, except for the broad peak close to the external cavity round trip frequency ν_{EC} which is marked in the figure. The bandwidth of the dynamics even comprises frequencies exceeding ν_{EC} by several hundreds of MHz. For frequencies beyond 4.5 GHz, the spectrum reveals a steep fall-off. However, the broad range of almost equal intensity distribution between about 500 MHz and 4.6 GHz is astonishing. It reveals the pronounced irregularity of the dynamics. We note that the frequency range can be extended to significantly exceed 10 GHz without losing the flatness of the rf spectrum. This can be simply done by reducing the delay time τ_{EC} and, therefore, shifting ν_{EC} to higher frequencies, which is accompanied by an increasing bandwidth of the dynamics [86]. However, detection becomes more challenging for increasing the bandwidth.

The spectral analysis of the intensity dynamics has revealed distinct differences for the SCR dynamics if compared to the LCR dynamics. The most relevant differences consist in the relevance of the delay time which dominates the dynamics in the SCR, while the relaxation oscillation frequency seems to be less important. Therefore, the dynamics in the SCR can give rise to very broadband evenly distributed intensity dynamics comprising frequencies well beyond the relaxation oscillation frequency. In contrast to this, the bandwidth of LCR dynamics is less pronounced, since the bandwidth of the dynamics is essentially limited by the relaxation oscillation frequency. Thus, we further investigate the striking differences between the SCR and LCR dynamics by comparison of the auto-correlation properties.

Auto-Correlation Properties

The auto-correlation functions of typical LCR and SCR dynamics of the SL with delayed optical feedback are depicted in Figure 3.23. In Figure 3.23 a), we present the temporal dependence of the auto-correlation coefficient for coherence collapse dynamics in the LCR which is scaled in units of the delay time τ_{EC}. In this case, the delay time is $\tau_{EC} = 2.9$ ns, the feedback ratio amounts to $r = 0.05$ and $p = 1.30$. The inset of panel a) illustrates a zoom into the auto-correlation function which exhibits recurrence of the auto-correlation for multiples of the delay time being characteristic for delay systems. The auto-correlation function exhibits a fast fall-off within $t_{e^{-1}} = 2\tau_{EC}$. This corresponds to an absolute decay time of $t_{e^{-1}} = 5.8$ ns. Additionally, we can identify that Σ_{corr} rapidly drops to almost zero between the peaks associated with τ_{EC} and that the recurrent increase of the magnitude of Σ_{corr} at $t \approx \pm 7\tau_{EC}$ is insufficient to

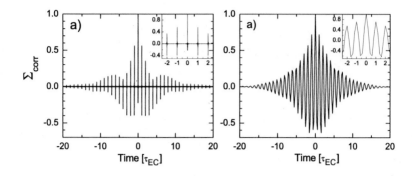

Fig. 3.23: Comparison of the auto-correlation properties of the intensity dynamics in the long cavity regime, depicted in panel a), and in the short cavity regime, illustrated in panel b). In panel a) $L_{EC} = 40$ cm, with $\tau_{EC} = 2.9$ ns, while in panel b) $L_{EC} = 3.9$ cm which corresponds to $\tau_{EC} = 260$ ps. The constantly kept parameters are: $p = 1.30$ and $r = 0.05$.

pass the $1/e$ bench mark. Both are indications that the dynamics operate well within the fully developed coherence collapse regime.

For comparison, Figure 3.23 b) presents the auto-correlation behavior of the broadband irregular SCR dynamics illustrated in the Figures 3.21 b) and 3.22 b). At this point, we note that the details of the auto-correlation function for the SCR dynamics are not fully resolved. This is due to the fact that the external cavity round trip frequency, which plays a crucial role for the SCR dynamics, amounts to $\tau_{EC} = 3.8$ GHz and, hence, it is close to the analog bandwidth of the oscilloscope being 4 GHz. Nevertheless, the auto-correlation function still provides meaningful information about the dynamics. The comparison of the auto-correlation properties of the SCR dynamics with that of the LCR dynamics reveals a faster fall-off of Σ_{corr} for the SCR dynamics. We estimate a range of Σ_{corr} of $t_{e^{-1}} = 4\tau_{EC}$ which corresponds to $t_{e^{-1}} \approx 1$ ns. This represents a very fast decay of the auto-correlation on absolute time scale. Unfortunately, we cannot precisely determine the first zero-crossing of the autocorrelation function to allow for estimation of the attractor dimension according to the Le Berre conjecture, as it was possible for larger delays for which detection bandwidth did not play a crucial role. Finally, we note that recent measurements with a state-of-the-art oscilloscope with 15 GHz analog bandwidth indeed revealed that the shape of the auto-correlation function in the SCR is also similar to that of the LCR. These measurements also showed that the actual range of the auto-correlation function of the fully resolved dynamics in the SCR for this pump parameter corresponds to $t_{e^{-1}} \approx 2\tau_{EC}$, similar to the case of the corresponding LCR dynamics in Figure 3.23 a).

In summary, the auto-correlation properties of SCR dynamics qualitatively agree to that of LCR dynamics, when all the operation parameters, except the delay time, are the same. This also holds for the qualitative dependence of the auto-correlation properties on the operation parameters, such as the pump parameter and the feedback strength. Therefore, it is in principal possible to further reduce the auto-correlation range by further increasing the pump parameter and optimizing the feedback strength, so that the magnitude of the auto-correlation drops below $1/e$ within $t_{-1} < 2\tau_{EC}$. Additionally, increasing the bandwidth of the dynamics by further reducing the delay time will also reduce the range of the auto-correlation function on an absolute time scale. However, the detailed characterization of temporal dynamics comprising such high bandwidth is still beyond the capabilities of presently available measurement technology. Nevertheless, recent developments of new generations of oscilloscopes indicate that this problem might be solved within the near future.

Spectral Properties

We have so far focused on characterization of the properties of the intensity dynamics in the SCR. In addition to this, the spectral properties can also provide insight into relevant fundamental mechanisms determining the dynamical properties. Our results for the LCR dynamics revealed that the intensity dynamics is directly connected to the spectral properties of the ECSLs. Since the intensity dynamics in the SCR already shows distinct differences with respect to the intensity dynamics in the LCR, it is an interesting question whether this interrelation holds for the corresponding spectral properties of the dynamics.

To gain insight into the spectral properties of feedback induced broadband chaotic dynamics, we present the optical spectra for a single mode SL (Optospeed). We note that the presented results for the single mode SL are in good qualitative agreement with the observed behavior for the spectral properties of the dominant longitudinal modes of multimode SLs operating in the SCR. The reason for this is that the coupling between the LMs of conventional multimode SLs is also weak for the considered range of parameters, so that multimode effect are not crucial. Here, we present results for the spectral properties of a single mode SL only for reasons of clarity.

In the experiment, the SL is subject to delayed optical feedback with a strength of $r = 0.04$, for which we measure a threshold reduction of $1 - I_{th,fb}/I_{th,sol} = 0.09$. The delay time amounts to $\tau_{EC} = 307\,\text{ps}$ corresponding to an external cavity round trip

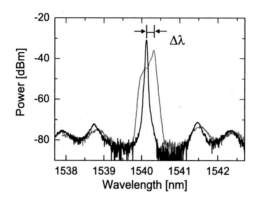

Fig. 3.24: Feedback induced linewidth broadening for chaotic dynamics of the IPAG DFB
semiconductor laser operating in the short cavity regime. The black line represents the spectrum of solitary laser, while the spectrum of the ECSL for chaotic
dynamics is represented in gray. The conditions are: $L_{EC} = 4.6$ cm, $\tau_{EC} = 307$ ps,
$p = 1.5$, and $r = 0.04$.

frequency of $\nu_{EC} = 3.25$ GHz. The SL is pumped at $p = 1.5$ and the feedback phase has
been adjusted for maximum bandwidth of the intensity dynamics for these conditions.
For these conditions, we find similar properties of the intensity dynamics as for the
dynamics presented in Figure 3.22 b). For these conditions, we study the spectral
emission properties of the SL system.

The gray line in Figure 3.24 represents the optical spectrum of the Optospeed DFB
SL for the specified operation conditions. For comparison, we provide the optical spectrum of the solitary SL without feedback, but for otherwise agreeing parameters. The
comparison of both spectra reveals that the linewidth of the emitted light substantially
broadens, while we also identify a red-shift of $\Delta\lambda = 195$ pm. This observation is in
agreements with the feedback induced linewidth broadening and red-shift which we
have identified for the LCR. However, a close look at the shape of the optical spectrum for the broadband irregular intensity dynamics reveals a subtle but characteristic
difference to the spectral properties of LCR dynamics. In addition to the red-shift of
the linewidth, we also find an asymmetric extension of the spectrum towards shorter
wavelengths, if compared to the center wavelength of the solitary laser mode. This
property reflects another peculiarity of the SCR dynamics, which we do not find for
LCR dynamics. In the LCR, the dynamics in inversion-frequency phase space, as we

have discussed for Figure 3.6, exhibits irregular pulsations in vicinity of the external cavity modes, while it continuously drifts towards higher gain. Eventually, the dynamics states gets close to an anti-mode where it hits the unstable manifold. This results in subsequent reinjection near the spectral position of the solitary SL mode. This reinjection process slightly differs for SCR dynamics in that the dynamics follow a well-confined reinjection channel guaranteeing continual reinjection of the dynamics into a region in phase space with an optical frequency beyond that of the solitary SL mode [87]. The well-confined injection channel is characteristic for dynamics in the SCR and results from the large spectral spacing of the external cavity modes.

In addition to the qualitative findings, the spectral properties even allow for estimation of some quantitative measures. First, we can calculate the frequency difference between the solitary SL mode and the maximum of the spectrum for broadband SCR dynamics. This frequency difference amounts to $\Delta\nu = 25.3\,\text{GHz}$. Since we know that the dynamics predominantly evolves towards the maximum gain mode, we can assume that the spectral position of the maximum gain mode roughly corresponds to the spectral position of the maximum of the optical spectrum for SCR dynamics. Thus, the red-shift allows us to experimentally estimate the extension of the corresponding fixed points structure in frequency space, being $\Delta\nu = 25.3\,\text{GHz}$. From this result, we can estimate the number of external cavity modes being involved in the dynamics. The number of external cavity modes is approximately given by the ratio between the frequency corresponding to the feedback induced red-shift divided by the external cavity round trip frequency: $\Delta\nu/\nu_{EC} \approx 8$. Interestingly, the estimated number of external cavity modes that are involved in the dynamics roughly agrees with the estimated dimension obtained from the Le Berre conjecture. However, we will determine attractor dimensions of SCR dynamics more rigorously in the following subsection, in which we present numerically obtained results for the properties of broadband chaotic dynamics generated in the SCR.

Modeling Results

In analogy to the numerical investigations of the LCR dynamics, in the following, we present numerically obtained results for modeling of broadband chaotic SCR dynamics. The procedures for computation of the time series and determination of the dynamics measures are the same as described for the LCR dynamics. Additionally, we introduce the Kolmogorov-Sinai entropy to characterize the dynamical properties of the corresponding attractors. In a first step, we adapt the parameters of the model to the experimental conditions, so that we find similar intensity dynamics. Then, we study the dependence of the Lyapunov exponents on the delay time, since the experiments

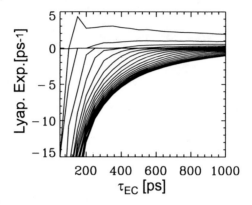

Fig. 3.25: The 20 largest Lyapunov exponents as a function of the delay time τ_{EC} for $\gamma = 10\,\text{ns}^{-1}$ and $p = 1.5$. Courtesy of R. Vicente.

revealed several differences between chaotic dynamics in the LCR and in the SCR. We can gain insight into the origin of these differences by studying the dynamical properties utilizing NLD measures.

Figure 3.25 depicts the dependence of the 20 largest Lyapunov exponents on the delay time τ_{EC} for moderate feedback $\gamma = 10\,\text{ns}^{-1}$ and pumping well above laser threshold $p = 1.5$. Additionally, we fix the optical feedback phase to $\Delta\Phi = \pi/2$ for which we find highest bandwidth of the dynamics. The results reveal that the number of positive Lyapunov exponents continuously decreases for shortening the delay time, when the dynamics evolves from the LCR into the SCR. In the SCR, for $\tau_{EC} < 800\,\text{ps}$, we identify a very interesting behavior of the magnitude of the largest Lyapunov exponent. The largest Lyapunov exponent steadily increases for reducing τ_{EC} until it reaches a maximum at $\tau_{EC} \approx 150\,\text{ps}$. For shorter delay times, the magnitude of the largest Lyapunov exponent rapidly decreases so that the dynamics becomes stable at about $\tau_{EC} \approx 100\,\text{ps}$. This suggests that the irregularity of the dynamics may exhibit a maximum within the SCR, while the geometric dimension of the attractor of the dynamics reduces for shorter cavities. Nevertheless, for similar conditions as for the experiment, for which the delay time was $\tau_{EC} \approx 300\,\text{ps}$ we find both, high-valued Lyapunov exponents and a high-dimensional chaotic attractor with a computed Kaplan-Yorke dimension of $d_{KY} \approx 8$, as it can be deduced from Figure 3.20. Interestingly, the computed dimension of the dynamics for these conditions agrees very well with the experimental estimate, which we have obtained from analysis of the optical properties.

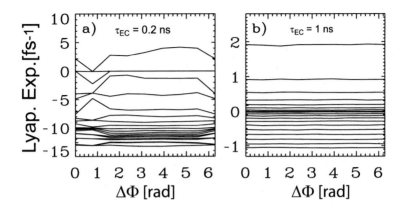

Fig. 3.26: The 20 largest Lyapunov exponents as a function of the feedback phase $\Delta\Phi$ for $\gamma = 10\,\text{ns}^{-1}$ and $p = 1.5$. Panel a) depicts the behavior for $\tau_{EC} = 0.2\,\text{ns}$, while in panel b) $\tau_{EC} = 1\,\text{ns}$. Courtesy of R. Vicente.

Next, we study the influence of the optical feedback phase $\Delta\Phi$ on the attractor of LCR and SCR dynamics to gain information about the relevance of this system parameter. Figure 3.26 depicts the 20 largest Lyapunov exponents as a function of $\Delta\Phi$ for two different delay times $\tau_{EC} = 0.2\,\text{ns}$ and $\tau_{EC} = 1\,\text{ns}$, respectively. We consider moderate pumping of $p = 1.5$ and set the feedback rate to $\gamma = 10\,\text{ns}^{-1}$. Panel a) discloses the dependence of the largest Lyapunov exponent on $\Delta\Phi$, as it was suggested by the experimentally observed sensitive dependence of the dynamics on $\Delta\Phi$. Even more, it is possible to recognize the transition between stable emission, periodic states and chaotic dynamics for carefully studying the Lyapunov spectra. In contrast to this, the magnitude of the largest Lyapunov exponent of the LCR dynamics only marginally changes for varying $\Delta\Phi$, as it can be seen in panel 3.26 b). As a consequence thereof, the LCR dynamics is almost insensitive on variations of the feedback phase $\Delta\Phi$. Nevertheless, the observed small sensitivity on $\Delta\Phi$ also implies that the feedback phase still influences the dynamics in the LCR. However, only sufficiently sensitive methods allow for detection of this marginal influence of $\Delta\Phi$ on the dynamical properties in the LCR [144].

Up to now, we have only discussed the geometric dimension of the chaotic attractor of the dynamics, but the presented results also reveal considerable characteristics for the dependence of the magnitude of the Lyapunov exponents on the relevant system parameters. This is particularly the case in the short cavity regime for which we find the maximum magnitude of the largest Lyapunov exponent. Additionally, the Lyapunov exponents of SCR dynamics sensitively depend on the feedback phase. These results suggest that solitary characterization of the dynamics in terms of dimensionality is probably insufficient, since it does not provide information about the dynamical properties of the underlying corresponding strange attractor.

The dynamical characteristics of a strange attractor can be expressed by a measure which can be derived from the Lyapunov spectrum. For this task, it is necessary to describe the statistics of the dynamics in the framework of information theory. Then, it is possible to express the "degree of chaos" of the dynamics of a system by a generalized concept of entropy for the state space dynamics. The corresponding measure is the Kolmogorov-Sinai entropy (h_{KS}), which is inversely proportional to the time interval over which the future evolution of the dynamics is predicted. More precisely, h_{KS} describes the rate of the average loss of information. The Kolmogorov-Sinai entropy of regular dynamics is zero, while it is positive infinity for a perfectly stochastic process. Hence, higher values for h_{KS} imply larger unpredictability of the future of the dynamics. The Kolmogorov-Sinai entropy can be calculated from the positive Lyapunov exponents using the so-called Pesin identity [142, 145]:

$$h_{KS} \;=\; \sum_{i|\lambda_i>0} \lambda_i \qquad (3.16)$$

The Pesin identity states that h_{KS} is identical to the sum of all the positive Lyapunov exponents. Using this identity, it is relatively easy to determine h_{KS}, once the Lyapunov spectrum has been calculated. Nevertheless, in a strict sense, the Pesin identity provides only an upper bound of the actual value of h_{KS}, but it has been established in numerous examples that the Pesin identity generally holds for dynamical systems.

We apply the concept of characterization of dynamical properties in terms of the Kolmogorov-Sinai entropy on the dynamics of the present ECSL system. Therefore, we compute the Lyapunov spectra for variation of the feedback strength, the delay time and the pump parameter. From the Lyapunov spectra in turn, we determine the Kolmogorov-Sinai entropy characterizing the dynamical properties of the attractor. The results are presented in Figure 3.27. The inset in Figure 3.27 a) illustrates the dependence of h_{KS} on the delay time for otherwise fixed parameters, i.e., $\gamma = 10\,\text{ns}^{-1}$ and $p = 1.5$. We find that h_{KS} exhibits a strong dependence for short delay times. At about $\tau_{EC} = 180\,\text{ps}$, the Kolmogorov-Sinai entropy reaches its maximum, while it slightly decreases for increasing delay times until it remains almost constant for large delay times, corresponding to operation of the ECSLs in the LCR. This behavior can be explained by the dependence of the magnitude of the Lyapunov exponents on the delay time. An increase of the delay induces a growth of h_{KS} because of an enhancing

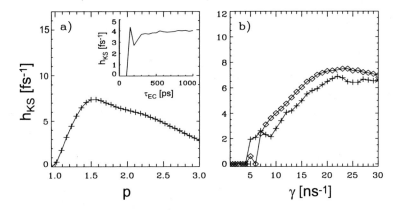

Fig. 3.27: Panel a) presents the Kolmogorov-Sinai entropy (h_{KS}) as a function of the pump parameter for $\tau_{EC} = 1$ ns. The inset in panel a) shows the Kolmogorov-Sinai entropy as a function of the delay time for fixed feedback rate of $\gamma = 10$ ns^{-1} and $p = 1.5$. Panel b) shows the dependence of h_{KS} on the feedback rate for $\tau_{EC} = 0.2$ ns (crosses) and $\tau_{EC} = 1$ ns (diamonds). Courtesy of R. Vicente.

number of positive Lyapunov exponents, as we have seen in Figure 3.25. However, as the Lyapunov exponents become smaller for increasing the delay, h_{KS} almost remains constant for long delay times.

To verify whether the Kolmogorov-Sinai entropy also exhibits a characteristic dependence on the operation parameter, we perform similar calculations for varying the pump parameter and the feedback strength, respectively. We note that in these calculations, we have chosen $\tau_{EC} = 1$ ns corresponding to operation of the ECSLs in the LCR. First, we investigate the dependence of h_{KS} on the pump parameter as it can be seen in Figure 3.27 a). The results disclose a well-defined maximum of h_{KS} at about $p = 1.5$. This pump parameter represents an optimal pump parameter with respect to maximization of the entropy of the dynamics. Certainly, the Kaplan-Yorke dimension is higher for $p \approx 2$ because of the larger number of positive Lyapunov exponents. Nevertheless, the Kolmogorov-Sinai entropy has its maximum at this point, because it essentially depends on the magnitude of the largest Lyapunov exponent which reaches its maximum for $p = 1.5$. Therefore, we also find this characteristic dependence of the Kolmogorov-Sinai entropy on the pump parameter in the SCR dynamics, in which h_{KS} also exhibits a maximum value for variation of the pump parameter.

We complement our investigation of the SCR by studying the dependence of h_{KS} on the feedback strength γ for $\tau_{EC} = 0.2\,\mathrm{ns}$ and $\tau_{EC} = 1\,\mathrm{ns}$, while we keep the pump parameter fixed at $p = 1.5$. The results are presented in Figure 3.27 b), which show that the Kolmogorov-Sinai entropy is slightly larger for the LCR dynamics than for the SCR dynamics. Nevertheless, the qualitative dependence of h_{KS} on γ agrees in both regimes. In the LCR as well as in the SCR, we find a weakly pronounced maximum of h_{ks} for moderate feedback rates of $\gamma \approx 20\,\mathrm{ns}^{-1}$.

In conclusion, the complementary numerical modeling of the broadband SCR dynamics which we have experimentally investigated has provided valuable information. Firstly, we have analyzed the Lyapunov spectra of the dynamics and verified that the dynamics is indeed of chaotic nature. Secondly, our results demonstrate that the entropy of the dynamics can also reach substantially high values for SCR dynamics, due to the fact that the largest Lyapunov exponent exhibits its maximum magnitude within the SCR. Thirdly, the Kolmogorov-Sinai entropy gives evidence for optimal values for the feedback strength and for the pump parameters. These optimal values correspond to standard operation conditions that can be easily achieved for conventional SLs which are subject to optical feedback. In summary, our results show that ECSLs which operate in the SCR can exhibit interesting high-dimensional broadband chaos with dynamical properties that can be excellently controlled and tuned within a wide range. For that reason, the ECSLs operating in the SCR regime represent very versatile and well-controllable experimental systems for studying high-dimensional broadband chaos. Even more, these excellent dynamical properties of short ECSLs systems offer high potential for applications which require well-controllable generation of high-dimensional broadband chaotic signals, such as for encrypted communication systems.

Finally, note that so far we have studied ECSLs which exhibit dynamical properties that can be modeled within the framework of the Lang-Kobayashi SL rate-equation approach. It has turned out that conventional SLs usually exhibit dynamics which can be described by the LK equations. However, we point out that this is not the case in general. Some of the ECSL systems can exhibit different, albeit very interesting, dynamical behavior. To remind one of this possibility, we briefly address in the following the topic about so-called "non-Lang-Kobayashi SL systems".

Non-Lang-Kobayashi Semiconductor Laser Systems

In this chapter, we have mainly focused on ECSLs for which L_{EC} was sufficiently longer than the semiconductor laser cavity L_{SL}. Comparison between experimental and numerically obtained results have turned out that most of the dynamics of these systems is captured by the Lang-Kobayashi SL rate-equation model. For that reason the LK equations have demonstrated their relevance since they can provide insight into dynamical properties that are presently experimentally not accessible, such as Lyapunov exponents, dimension analysis and entropy measures. However, we have also seen that the derivation of the LK equations is based on several assumptions and simplifications which are not fulfilled in general. Two main assumptions of this mean-field model are the consideration of only one (single) longitudinal SL mode and the neglect of the longitudinal extension of the semiconductor laser cavity. However, in some SLs further modes and spatial dependencies can become important for the dynamics. This particularly can be found e.g. for some modern types of SL which currently attract considerable attention, like transversally two-dimensional Vertical-Cavity Surface-Emitting SLs and one-dimensional Broad-Area SLs. Such effects can also arise in rather conventional edge emitting SL. Then, it might be necessary to capture modal dynamics which is not considered in the basic LK model. In Subsection 3.2.2, we have discussed alternative models which can be applied in one or the other situation. Unfortunately, all of the models exhibited pros and cons since their derivation is also either based on assumptions and simplifications which do not apply in general, or the models are very sophisticated hampering physical interpretation of the results. Nevertheless, these models have considerably contributed to today's understanding of multimode dynamics in SLs, which we have not considered so far. The reason for this was that the dynamics of conventional SLs typically exhibits only weakly pronounced multimode effects. Nevertheless, interest in multimode effects is growing. This is due to the fact that modern technology allows for production of larger SL devices which can provide higher output power and which are required for many applications, e.g., in material processing or spectroscopy. Unfortunately, increase in spatial extension is often paid for degradation of the spectral emission properties because of multimode emission. Therefore, analysis and understanding of multimode effects in SL dynamics is essential for the functionality of many applications. Additionally, such experiments can provide fundamental insight into properties of pronounced multimode systems. In the following Chapter, we present such an interesting SL system which exhibits distinct multimode emission dynamics, well beyond the LK description. We will demonstrate that the nonlinear dynamical properties of the designed SL system exhibits potential for novel applications of SLs.

3.5 Concluding Summary

In this chapter we have presented a comprehensive analysis of the dynamical properties of SLs which are subject to time-delayed optical feedback and therefore emitting chaotically. We have applied state-of-the-art measurement technology for experimentally studying the dynamics of ECSLs operating in two different dynamical regimes,

Properties	Long Cavity Regime	Short Cavity Regime
system frequencies	$\nu_{EC} \ll \nu_{RO}$	$\nu_{EC} \gtrsim \nu_{RO}$
sensitivity on $\Delta\Phi$	not noticeable	yes
control of dynamics	very good	excellent
robustness	very good	very good
compactness	no	yes
continuous rf spectrum	no	yes, up to ν_{EC}
rf bandwidth	$\nu_{rf,max} \sim \nu_{RO}$ typ. $\nu_{r,max} < 10\,\mathrm{GHz}$	$\nu_{rf,max} \sim \nu_{EC}$ $\nu_{rf,max} \gg 10\,\mathrm{GHz}$ poss.
correlation time	$\gg 1\,\mathrm{ns}$	$< 1\,\mathrm{ns}$
$d_{KY}(\sharp_{\lambda_i > 0})$	$\gg 10$	~ 10
$\lambda_{i,max}$	$\exists$ max.	$\exists$ max.
$h_{KS}(\Sigma_{\lambda_i > 0})$	saturates	$h_{KS,max}$ in SCR

Table 3.2: Characteristics of long and short cavity regime dynamics.

the long and the short cavity regime. Our results demonstrate that in both regimes the dynamics can exhibit high-dimensional broadband chaos in conjunction with well-pronounced spectral dynamics. However, the dynamical properties and dependencies in both regimes exhibit different behavior. This is mainly due to the different physical mechanism dominating the dynamics. In the LCR, we find that the relaxation oscillation frequency represents the fundamental physical mechanism domination the dynamics. In contrast to this, we identify that in the SCR the delayed feedback plays the major role for the dynamics. We have studied the peculiarities of the dynamics in each of the two regimes for which we also have applied complementary numerical modeling. This approach proved to be very beneficial, since based on the well-agreeing results for the experiments and modeling, the numerical modeling allowed for characterization of the geometrical and the dynamical properties of the chaotic attractors. In conclusion, both the LCR and the SCR regimes reveal pronounced high-dimensional broadband chaotic dynamics which are experimentally well-controllable. Even more, we have demonstrated that it is possible to manipulate the dynamical properties in both regimes. In that sense, NLD can be utilized for the controlled manipulation of the emission properties of chaotically emitting SL. This applies to both, to the properties of the intensity dynamics as well as to the properties of the spectral dynamics, since they turn out to be strongly interrelated. This quality of ECSLs systems is very attractive from the application-oriented point of view, since it allows for tailoring of the emission properties of SLs according to the demands of the technical application. In particular, it offers high potential for novel applications which are based on the properties of chaotic light, such as in encrypted communications or for Chaotic Light Detection and Ranging (CLIDAR).

Since we are also interested in this fascinating possibility of harnessing NLD for new functionalities, we summarize our main results in Table 3.2. The table provides an overview over practical aspects and dynamical properties for both regimes of operation, the long and the short cavity regime. Therefore, it allows for direct comparison of the characteristics of both dynamical regimes, which we will refer to later on in this thesis.

Chapter 4

Tailoring the Emission Properties of Semiconductor Laser Systems

In the previous chapter, we have discussed the influence of delayed optical feedback on the emission properties of SLs. We have demonstrated that the combination of the delay and the inherently strong nonlinearity of SLs are ideal premises for the occurrence of high-dimensional chaotic intensity dynamics in conjunction with pronounced spectral dynamics. Our experiments have demonstrated that the dynamical properties of these SL systems (ECSLs) sensitively depend on the operation conditions, such as the feedback conditions and the pump parameter; and device characteristic parameters, e.g., the α-parameter and the properties of the relaxation oscillations. We have shown that the emission properties of SLs can be controlled by variation of the system parameters. This allows for precise adjustment of dynamical states between stable emission and broadband chaos. As a consequence thereof, the according spectral dynamics, determining the coherence properties of the SLs, can also be tuned. In that sense, application of NLD offers potential for tailoring the coherence properties of SLs, in terms of the dynamically accessible range of the particular SL system.

Almost four decades years ago, it has been shown that the influence of optical feedback can be ambivalent. It has been found, that optical feedback not only can induce instabilities, but that it can also exhibit stabilizing effects on dynamical systems. In recent years, this property of delayed feedback has been successfully applied for stabilization of the emission of SLs [75–77], while much effort has been aimed for avoidance of feedback induced linewidth broadening. Motivation was given because linewidth broadening often degrades the performance of practical devices. Therefore, studies on the influence of feedback induced intensity dynamics on the coherence properties of SLs are rare. The reported works mainly focus on conventional regimes of operation [136–138], i.e., operation within the long cavity regime and presence of weak to moderate feedback. Up to now, detailed investigations of the feasibility of the controlled manipulation of the emission properties by harnessing the nonlinear dynamical properties of SLs are still lacking. The possibility for specifically reducing the coherence length of SLs is particularly interesting since it offers potential for realization of novel applications for which (in-)coherence properties are of importance.

Lately, modern technical applications have been proposed which are based on bright incoherent light sources with good beam properties, e.g., coherence tomography [146] and chaotic light detection and ranging (CLIDAR) [71]. These applications raise high demands on the emission properties of the light sources, since for conventional light sources brightness and incoherence often goes along with reduced quality of the beam properties, and vice versa. Currently, there are only few incoherent light sources available which are based on different approaches. One prominent approach utilizes nonlinear fibers and high power light sources for generation of supercontinuum light. Functionality of supercontinuum light sources is based on nonlinear interactions between the intense light and the material of the fiber for enhancement of the spectral bandwidth of the injected light. Continuum generation is the result of multiple nonlinear phenomena such as stimulated Raman scattering, self-phase and cross-phase modulations, four wave mixing, high-order soliton formation and parametric mixing through modal phase matching in the case of multimode optical fibers [147, 148]. Supercontinuum light sources can provide light comprising a spectral bandwidth of up to the order of one micrometer while reaching output powers between tens and hundreds of mW, depending on the configurations and usage of optical amplifiers. On the one hand, the nonlinear properties of the fiber need to be maximized for optimization of performance. This can be done by using appropriate materials or by application of specially structured fibers, such as novel photonic-crystal fibers, or tapered fibers. On the other hand, the intensity and the spectral bandwidth of the injected light influences the bandwidth of the output. Therefore, specialized light sources are required as injectors, e.g., ps-pulse solid state or fiber lasers, or amplified spontaneous emission sources. These devices represent sophisticated light sources which can be quite expensive and impractical. Super-luminescence diodes provide compact and cost-effective alternatives. These type of semiconductor diodes are optimized for broadband emission by tailoring the cavity of the diode. Therefore, the emission properties of these light sources typically represent a trade-off between bandwidth and output power. Currently available super-luminescence diodes offer spectral bandwidths in the range between $\sim 15\,\mathrm{nm}$ and $\sim 120\,\mathrm{nm}$ with emission powers between $\sim 25\,\mathrm{mW}$ and $\sim 0.5\,\mathrm{mW}$, depending on the gain material and structural design. The upper bound of the emission bandwidth is defined by the spectral bandwidth of the material gain which can be exploited in the absence of a resonator, unfortunately the absence of a resonator drastically reduces the output power, and usually results in loss of lasing. For increasing the output power, the light is typically amplified by application of a (weak) resonator, which in turn reduces the bandwidth of the emitted light. Meanwhile, various resonator geometries, have been proposed for optimization of the performance of super-luminescence diodes. New generations of these diodes offer higher output power for substantially extended gain regions, but this approach also reduces the qualities of the beam [149]. A potential alternative consists in utilization of newly developed gain structures such as quantum dot structures which offer substantially higher bandwidths of the material gain than conventional SLs. However, accurate production of these structures is challenging and fabrication processes have just reached a matured state allowing for reliable and cost-effective production. Therefore, currently, it is an open question whether this approach will be fruitful. All in all, the presently available incoherent light sources exhibit advantages, but also specific drawbacks which reduce system performance and restrict applicability. Thus, alternative incoherent light sources are desired and welcome.

In this chapter, we will address the question whether the nonlinear dynamical properties of SL with delayed optical feedback can be harnessed for realization of such an alternative incoherent light source. In this context, we benefit from the insight we have gained from the detailed analysis of the dynamical properties of ECSLs. We apply the obtained knowledge of the interrelation between intensity dynamics and spectral dynamics to tailor a unique SL that exhibits unusual spectral broadband emission dynamics. We study the dynamical properties of the system and find distinct dynamical properties which are clearly beyond description within the framework of LK modeling. In particular, we reveal distinct differences to the short cavity regime of conventional LK SL systems reported so far. The most prominent difference is the onset of chaotic intensity dynamics in conjunction with pronounced multimode dynamics of high optical bandwidth exceeding $\sim 7\,\mathrm{nm}$, therefore comprising more than 100 lasing longitudinal modes (LM). On the one hand, the presented system represents an excellent nonlinear dynamical model-system offering well-controllable generation of distinct multimode dynamics. On the other hand, we demonstrate that the distinguished nonlinear dynamical properties allow for controlled adjustment of the coherence length in a wide range between $\sim 130\,\mu\mathrm{m}$ and $\sim 8\,\mathrm{m}$. This verifies that NLD indeed can be successfully harnessed for the controlled manipulation of the emission properties of SLs, allowing for drastic reduction of the coherence length. Hence, SLs with optical feedback can be implemented in novel measurement technology in which (in-)coherence properties are of importance.

4.1 Design of an Incoherent Light Source

To reduce the coherence properties of a light source, it is essential to increase the spectral bandwidth of the source. In Chapter 3.3.2, we have demonstrated that delayed optical feedback can induce dynamics in SLs which manifest itself in broadening of the linewidth of the emitted light. This effect can be utilized for increasing the spectral bandwidth of the emitted light and reducing its coherence length. We have demonstrated that the feedback induced linewidth broadening is proportional to the intrinsic nonlinearity of the SL, the α-parameter or linewidth enhancement factor. Additionally, the linewidth broadening is directly proportional to the feedback strength, as it can be deduced from Eqs. 3.6-3.7. Besides these two key parameters, further parameters and conditions entering the dynamical properties are also relevant for the coherence properties of ECSLs.

So far, we have mainly focused on the emission properties of LK SL systems with external cavities L_{EC} being sufficiently longer than the semiconductor laser cavity L_{SL} [86, 88, 98, 117, 119]. This condition comprises operation in the long cavity regime and the short cavity regime, discussed so far. For operation in these regimes, we have demonstrated that linewidth broadening allows for reduction of the coherence lengths of a SL typically down to the order of 5 millimeters. However, coherence lengths of the order of 5 millimeters are quite long with respect to modern high-resolution metrology applications. For these applications, the coherence length roughly defines the measurement resolution. Thus, in terms of practical applications further reduction of the coherence length is strongly desired. Our goal is to drastically reduce the coherence properties of SLs by application of delayed optical feedback and by harnessing the nonlinear dynamical properties of such SL systems. Our strategy to achieve this goal comprises several potentialities for optimization: selection of the SL device, external cavity design, and operation parameters.

Multimode lasers represent very promising candidates, since each of the lasing longitudinal laser modes (LMs) contributes to the optical bandwidth of the emitted light. This has been verified in Chapter 3.3.2. Therefore, in the following, we focus on the realization of a tailored, pronounced multimode SL system with numerous strongly interacting LMs being involved in the emission dynamics. The following strategies will support the realization of such a system:

- Increase of the number of LMs is desired for enhancing the effective optical bandwidth. This can be achieved by:
  - selection of a SL with high bandwidth of gain
  - reduction of the LM spacing of the SL

- Enhancement of the coupling between the LMs is essential, which can be supported by:
  - selection of a SL with high α-parameter
  - application of moderate feedback

- realization of comparable lengths of both cavities for enhancement of the gain-coupling between adjacent LMs and reduction of the number of external cavity modes (ECMs) between adjacent LMs

- realization of a resonance condition between both cavities for enhancement of the coupling between the LMs and the ECMs

Realization of these strategies inevitably introduces system properties which are beyond description within the frame of LK modeling. In particular, implementation of the resonance condition for comparably long cavities is beyond capabilities of currently available LK models. However, besides our interest in the dynamical properties of such a system aiming for technical applications, the resulting dynamical properties are highly interesting from the fundamental point of view, since such systems have not been studied so far. Hence, realization and subsequent analysis of such a SL system is tempting from both, the fundamental NLD point of view and the application-oriented point of view.

4.2 Short Resonant Cavities Semiconductor Laser System

In this section, we will demonstrate that SL systems with properties beyond validity of the conventional LK description can reveal interesting dynamical phenomena including distinct multimode emission dynamics. Therefore, these systems can provide essential insight into the properties of complex nonlinear multimode systems in general. In addition, the intriguing characteristics of the emission properties of these systems can be harnessed for realization of novel technical applications which are based on utilization of chaotic light. In this context, we present a distinguished multimode ECSL system which clearly violates fundamental assumptions of LK modeling, i.e., single LM emission and neglect of the longitudinal extension of the semiconductor laser cavity. In contrast to conventional LK systems, the lengths of the external cavity L_{EC} and the length of the semiconductor laser cavity L_{SL} are of comparable size. For these conditions, the longitudinal extension of the SL cavity becomes important for the dynamics. This becomes evident when we additionally introduce resonance conditions between the two cavities to enhance the coupling between the LMs. As we will demonstrate, the combination of both features can trigger excitation of numerous strongly interacting LMs manifesting in extraordinary high optical bandwidth, if compared to conventional ECSL systems. Because of this striking difference, main focus of this work is dedicated to the emergence and analysis of the fascinating multimode dynamics exhibited by this SL system.

The organization of the chapter is as follows. In Subsection 4.2.1, we introduce the resonant short cavities SL system and describe the experimental setup which we utilize for characterization of its dynamics. In Subsection 4.2.2, we demonstrate how coupling between the LMs can be distinctly enhanced for resonant cavity conditions. In Subsection 4.2.3, we study the influence of the relevant system parameters on the dynamics, which are the pump parameter and the feedback phase. For variation of both parameters, we find a cyclic scenario in which the dynamics evolve from stable emission, via periodic states, to chaos, and subsequently again to stable emission. We show that for high pump parameters we are able to achieve broadband chaotic intensity dynamics for which the number of LMs being involved in this dynamics can easily exceed 100. We analyze the emergence of this extraordinary optical broadband dynamics disclosing the strong interrelation between intensity and spectral dynamics. Analysis of the coherence properties of this dynamics elucidates the high potential of this light source in terms of modern technical applications. Consequently, in Subsection 4.2.5, we study this extraordinary broadband dynamics in detail in spectrally resolved measurements to gain deeper insight into the processes underlying the pronounced multimode emission. We discuss the spectral dynamics and motivate for the development of new models which can unveil the essential mechanisms determining the characteristic dynamics of this system.

4.2.1   Experimental Setup

In the following, we introduce the resonant short cavities SL system and illustrate the experimental setup which we apply for analysis of the dynamical characteristics of the system. Figure 4.1 presents a schematic of the experimental setup. In the figure, the SL system is highlighted by a gray box which is located in the upper half of the figure, while the detection branch is mainly sketched in the lower half of the figure. The central device of the system is a ridge waveguide SL (FBH) which emits at a center wavelength of 785 nm. The SL has been selected according to a small spectral spacing of the LMs and a broad, flat gain profile. The length of the laser cavity of the FBH SL is $L_{SL} = 1.6$ mm and the effective refractive index of the gain material is $n = 3.7$. Consequently, the longitudinal mode spacing corresponds to 25.3 GHz, which is also the round trip frequency of the light in the SL cavity $\nu_{SL} = 25.3$ GHz. To guarantee well-defined operation conditions, the laser is pumped by an ultra-low-noise DC-current source and its temperature is set at 22.2° C and stabilized to better than 0.01 K. The maximum output power of the SL is 109 mW at the upper limit of the pump current

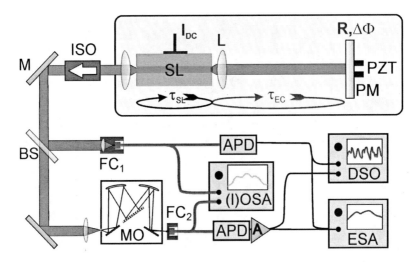

Fig. 4.1: Scheme of the experimental setup of the semiconductor laser feedback system.

being $I_{DC} = 150$ mA. For these conditions the maximum of the relaxation oscillation frequency of the carrier-photon system, which represents a relevant frequency for the dynamics of the FBH SL system, has been determined to be $\nu_{RO,max} = 3.3$ GHz. Furthermore, we have measured the linewidth enhancement factor which expresses the strength of the nonlinearity in the SL to be $\alpha = 2.0 \pm 0.2$. This has been done by applying the method proposed by Henning and Collins [61].

In the experiments, the light emitted from the front (right) facet of the SL is collimated by a lens (L) and propagates towards a partially transparent mirror (PM) with reflectivity R. A part of the light is reflected from the PM and is reinjected into the SL after the delay time $\tau_{EC} = 2 L_{EC}/c$, and with the phase difference $\Delta\Phi = \Phi(t) - \Phi(t - \tau_{EC})$. Here, L_{EC} denotes the length of the external cavity, and c represents the speed of light in air. In the experiments, the ratio between L_{EC} and the optical length of the semiconductor laser cavity $L_{SL,opt} = n\, L_{SL}$ has been chosen between 2 and 5. This corresponds to external cavity round trip frequencies of 5.1 GHz $\leq \nu_{EC} \leq 12.7$ GHz. For such short external cavity lengths, the short cavity regime (SCR) requirement [86, 115] is always fulfilled, since the external cavity round trip frequency $1/\tau_{EC}$ is always sufficiently larger than $\nu_{RO,max} = 3.3$ GHz. For these conditions, the key parameters of the system determining the dynamical behavior are the delay time τ_{EC}, the pump current I_{DC}, the feedback ratio r, the feedback phase $\Delta\Phi$, and the ratio between the length of both cavities M, with $M = L_{EC}/L_{SL,opt}$. The feedback ratio r describes the strength of the feedback and is defined as the ratio between the power of the light effectively coupled back into the SL and the power emitted at the front facet of the SL, $r = P_{fb}/P_{out}$. The control parameters can be varied by changing L_{EC}, exchanging the PM by one with different reflectivity R, by varying I_{DC} and by shifting the PM on sub-wavelength scale with a piezo-electric transducer (PZT). The light emitted from the rear facet of the SL is sent to the detection branch which is isolated by an optical isolator (ISO) to shield the SL system from unwanted feedback from the detection branch. The light is divided at the non-polarizing 50/50 beam splitter (BS). One part of the signal is coupled into a fiber at the fiber coupler (FC$_1$) and further subdivided for analysis of the emission properties. One fraction of the light is detected by a 12 GHz photodetector (APD), whose electric output is monitored on a digital storage oscilloscope (DSO) with an analog bandwidth of 4 GHz (at a sampling rate of 20 G samples/s), and an rf spectrum analyzer (ESA) with 21 GHz bandwidth. The other fraction of the light can be either analyzed utilizing an optical grating spectrum analyzer with a resolution of 24 GHz (OSA) or by an interferometric spectrum analyzer with a resolution of up to 4 GHz (IOSA, Advantest Q8341). To gain insight into the properties of the dynamics generated in different optical spectral regions, the part of the light which passes the BS is coupled to a grating monochromator (MO) where it is spectrally filtered. The filtered light passing through the MO is coupled into a fiber at FC$_2$ for detection. The optical bandwidth (3 dB) of the filtered light amounts to approximately 170 GHz, which is equivalent to an interval comprising 7 LMs of the SL. The center wavelength of the spectral filter can be tuned over the entire emission spectrum of the SL system. In analogy to the detection of the total intensity dynamics, the detection apparatus for the spectrally filtered dynamics is identical, except for an additional rf amplifier (A) inserted after the APD accounting for the reduced intensity caused by spectral filtering and additional losses due to lower coupling efficiency at FC$_2$. In the following, we refer to the corresponding measurements as spectrally resolved measurements, since the spectral bandwidth of the filtered dynamics is sufficiently smaller than the maximum optical bandwidth of the total emission dynamics.

Before we experimentally investigate the characteristic dynamics of the SL system, we motivate for a particular choice of external cavity lengths. More precisely, we introduce resonant coupling between the semiconductor laser cavity and the external cavity so that the longitudinal modes (LMs) of the SL resonantly couple to the external cavity modes (ECMs). Our goal is to enhance the coupling between the LMs to achieve pronounced multimode dynamics comprising high optical bandwidth. Subsequently, we experimentally analyze the total intensity dynamics of the SL system in dependence of the pump current and the feedback phase $\Delta\Phi$.

4.2.2 Resonant Cavities Condition

The coupling conditions between the semiconductor laser cavity and the external cavity are of crucial importance for the dynamical properties of multimode SLs with optical feedback. This is due to the fact that the coupling influences the interactions between the lasing modes which, in turn, determine the dynamics of the system. In similarity to compound cavity lasers [150, 151], resonance conditions between the cavities in this double-cavity system represent distinguished coupling conditions. This is particularly the case when the lengths of the resonators are of comparable size. Since we are interested in pronounced multimode dynamics, we have experimentally studied the possibility of substantial enhancement of the coupling between adjacent LMs by realizing resonance conditions.

In the experiment, the SL operates at $I_{DC} = 2.6 I_{th,sol}$ well above the solitary laser threshold current $I_{th,sol} = 45.72\,\mathrm{mA}$. It is convenient to express the pump level in terms of the pump parameter (p) which is defined by the ratio $p = I_{DC}/I_{th,sol}$. The SL is subject to moderate feedback with $r = 0.16$ inducing a threshold reduction of $\Delta r = -0.069$. We adjust the ratio between the length of the external cavity and the SL cavity $M = L_{EC}/L_{SL,opt}$ to $M \approx 3$. Then, we slightly vary L_{EC} around the resonance condition and monitor the effect on the optical spectrum of the emission. The results are presented in Figure 4.2. The optical spectrum for $M = 3.15$ presented in Figure 4.2 a) reveals several broad peaks. These peaks correspond to groups of LMs which meet the constructive interference condition between the emitted light and the feedback. The excited LMs within these groups are subject to feedback with similar $\Delta\Phi$, while the phase difference between these groups corresponds to multiples of 2π. When approaching the resonance condition, the phase variation over the spectral range which is induced by the detuning between the cavities becomes smaller. This crucially influences the spectral emission properties which can be deduced from comparison of Figures 4.2 a) and b). On the one hand, approaching the resonance condition we find an increasing number of excited modes within each of the groups of LMs, manifesting in broader peaks. On the other hand, Figure 4.2 b) reveals that the spacing between the groups of lasing LMs increases, because of the smaller detuning between the cavities. For the resonance condition set to $M \approx 3.00$ and $\Delta\Phi$ adjusted for constructive interference condition, we are able to achieve conspicuous broadband emission comprising numerous lasing LMs, which is clearly demonstrated in Figure 4.2 c). Indeed, the resonant coupling scheme turns out to be very efficient. Further tuning the cavity ratio to $M = 2.97$, which is presented in Figure 4.2 d), we detune away from the resonance condition and observe similar but inverse behavior as for approaching the resonance condition. Hence, best coupling conditions are accomplished when the round trip frequencies of both cavities, the semiconductor laser cavity $(\nu_{SL,opt})$ and the exter-

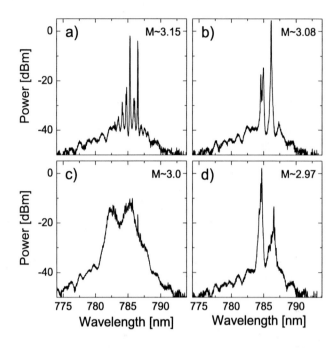

Fig. 4.2: Optical spectra for variation of L_{EC} around the resonance condition M=3. Longitudinal modes with similar $\Delta\Phi$ and which fulfill constructive interference conditions are excited. Approaching the resonance condition depicted in panel c), more and more LMs strongly couple and are excited. In panel a) the cavity ratio is $M \simeq 3.15$, in b) $M \simeq 3.08$, in c) $M \simeq 3.00$ and in d) $M \simeq 2.97$, respectively. Other conditions are $r = 0.16$ and $p = 2.6$.

nal cavity (ν_{EC}), fulfill an integer resonance condition so that $\nu_{SL,opt}/\nu_{EC} = M$, with M being an integer number between 2 and 5. For this prerequisite, both resonators strongly couple and adjacent LMs are equally supported in the gain medium (semiconductor laser cavity) allowing for substantial coupling between LMs. Additionally, the feedback phase $\Delta\Phi$ is well-defined in the SL system for integer resonance conditions, since each of the LMs is subject to feedback with the same value of $\Delta\Phi$. Depending on the gain profile, numerous LMs may be excited in the SL system offering potential for dynamics with considerable optical bandwidth.

Fractional ratios of $\nu_{SL,opt}/\nu_{EC}$, fulfill similar, although weaker, resonance conditions.

In particular, half-integer resonance conditions fulfilling $\nu_{SL,opt}/\nu_{EC} = M$ with $M = (2N + 1)/2$ and N being an integer between 2 and 5 represent comparatively good coupling conditions. For half-integer resonance conditions, every next but one LM is subject to the same $\Delta\Phi$, while each of the LMs in between are subject to the π-shifted phase value $\Delta\Phi + \pi$. Accordingly, for $\Delta\Phi = N\pi$, one of these two groups of LMs is supported in the gain medium in a similar manner as for the integer resonance condition, while the LMs in between are considerably less supported. Therefore, in case of half-integer resonance conditions, the mode spacing between every next but other mode is of importance for the emission properties. The corresponding frequency can be considered as mode spacing frequency of a resonator of half the length of the original semiconductor laser cavity, $\nu'_{SL,opt} = 2\,\nu_{SL,opt}$. For this frequency, in turn, an integer resonance condition is accomplished $M' = \nu'_{SL,opt}/\nu_{EC}$, with $M' = 2N$. Intuitively, this situation differs from the original integer resonance condition only in the sense that the frequency spacing of supported modes is larger, which reduces the coupling strength between the supported modes. To verify these considerations, we perform experiments in which we vary M in the range between 2 and 5, while the other operation parameters are kept constant. In fact, for sufficiently strong feedback, we find similarly good coupling conditions for half-integer resonance conditions of $M = 2.5, 3.5$ and 4.5 as for integer resonance conditions of $M = 2, 3, 4$ and 5. This qualitative agreement is demonstrated in Figure 4.3 in which we depict the corresponding optical spectra for $M = 3$, in Figure 4.3 a), and $M = 2.5$, in Figure 4.3 b). The figure discloses that for the integer and for the half-integer resonance condition the LMs efficiently couple manifesting itself in pronounced optical broadband emission. A high-resolution measurement using the IOSA, which offers a resolution of better than 4 GHz, can provide deeper insight into the spectral emission properties for optical broadband emission, since it allows for fully resolving the LMs. The result of such a measurement is depicted in Figure 4.4. The experimental conditions correspond to that of Figure 4.3 b) for $M = 2.5$. Figure 4.4 a) presents a seemingly noisy spectrum comprising the whole spectral bandwidth of the emission. We point out, that the spectrum is very stable on a time scale of several minutes. Hence, the spectrum only appears to be noisy. This property can be attributed to the high number of LMs being involved in the emission dynamics, which can be fully resolved by the interferometric optical spectrum analyzer. This becomes clear from a closer look at the spectrum presented in Figure 4.4 b), which shows a 1 nm zoom out of the central region of the spectrum presented in a). The zoom shows 19 peaks within this 1 nm interval corresponding to a spacing of 53 pm. This is exactly the LM spacing. Since the optical spectra for both resonance conditions presented in Figure 4.3 exhibit laser emission within a bandwidth of exceeding 7 nm, the high resolution measurement reveals that more than 130 LMs are involved in the emission. This represents a remarkably high number of LMs, if compared to the re-

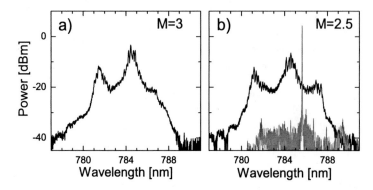

Fig. 4.3: Comparison of the spectral emission properties for integer cavity resonance condition $M = 3$, depicted in panel a), and half-integer resonance condition $M = 2.5$, shown in panel b). The feedback phase has been adjusted for maximum optical bandwidth. The feedback ratio amounts to $r = 0.16$ and pump parameter is $p = 2.9$. The gray line in Figure b) depicts the optical spectrum of the solitary laser without optical feedback.

ported multimode behavior of conventional non-resonant feedback SL systems. The comparison with the optical spectrum for singlemode operation in absence of optical feedback, which is represented by the gray line in Figure 4.3 b), highlights the distinct multimode emission. In analogy to recent experiments on solitary multimode SLs which revealed dynamically induced switching between interacting LMs [152, 153], the observed pronounced multimode emission properties immediately raise the question if the origin of the fascinating high optical bandwidth is of dynamical nature.

The anticipated answer is "yes". Accordingly, in the next subsection we analyze the dynamics of the SL system in detail and prove that for resonant coupling we are able to achieve chaotic intensity dynamics with an extraordinarily high optical bandwidth of several nanometers. For the following experiments, we choose a ratio of $M = 2.5$, since the larger spectral spacing of the supported modes facilitates better control over the dynamics. This property is especially helpful when studying the onset and the emergence of dynamics scenarios. We point out that the results we obtain for the dynamics for $M = 2.5$ agree to those for $M = 3$, except for some quantitative differences with respect to the dependence of the control parameters.

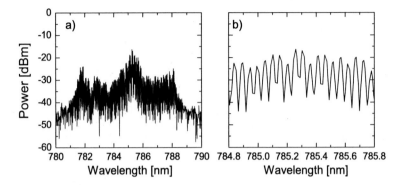

Fig. 4.4: High-resolution measurement of the optical properties for the same conditions as for the measurement of Figure 4.3 b). Full spectral bandwidth is presented in panel a), while panel b) presents a 1 nm zoom of the center region of the spectrum illustrated in a).

4.2.3 Scenarios of the Dynamics of the System

In the previous subsection, we have motivated our choice of the external cavity length being $L_{EC} = 2.5\,L_{SL,opt}$. With this condition, we have met two distinct fundamental preconditions with respect to the dynamics of the system. Firstly, the system fulfills the SCR requirement $\nu_{EC}/\nu_{RO} \gtrsim 1$ for the accessible range of parameters. Therefore, the resonant short cavities SL system is still a delay system being mathematically infinite dimensional. This offers potential for generation of high-dimensional dynamics. Secondly, another fundamental precondition is realized by selection of comparable round trip times inside the short external cavity, $\tau_{EC} \approx 100\,\mathrm{ps}$, and in the semiconductor laser cavity, being $\tau_{SL} = 2L_{SL,opt}/c \approx 40\,\mathrm{ps}$. For adjusted resonance conditions, this property allows for strong coupling of the LMs, because the LMs are only separated by very few (2-5) external cavity modes. Therefore, the fixed point structure of the system will be rather a global one and fundamentally different if compared to conventional multimode LK systems. In conventional LK systems, we typically find individual fixed point structures (LK ellipse) which are associated with the spectrally larger spaced LMs. Because of the wide spacing of the LM, the fixed point structures of adjacent LMs usually do not overlap. Hence, coupling between the LMs is not very strong in conventional LK systems.

In the following, we analyze the dynamical properties of the resonant short cavities

SL system and identify the pump current I_{DC} and the feedback phase $\Delta\Phi$ as major control parameters. Henceforth, we study their influence on the dynamical properties of the SL system.

Influence of the Pump Parameter

In this subsection, we investigate the dependence of the dynamics on the pump current. In the experiment, we set the feedback ratio to $r = 0.16$, inducing a threshold reduction of $\Delta r = -0.069$. For these conditions, we increase the pump parameter from $p = 0$ to $p \approx 3$.

Starting at $p = 0$, we find the onset of dynamics for $p = 0.93$. For such small values of the pump parameter, the dynamics consists of slow intensity fluctuations comprising frequencies of up to several hundred MHz. However, the intensity fluctuations are not equally distributed, but cluster around a center frequency giving rise to a broad peak in the rf spectrum. The peak frequency increases as the pump parameter is increased. The peak frequency of the slow intensity fluctuation is not always present. Instead, we find a (cyclic) scenario for increasing pump parameter on a scale of $\Delta p \approx 0.14$. In this scenario, the dynamics evolves from stable emission to slow intensity fluctuations with a center frequency of a few hundred MHz, until the dynamics suddenly disappears and the emission becomes stable again. Nevertheless, the average peak frequency of the dynamics increases continuously for incrementing pump current. For this feedback ratio, approximately 2 cycles of this scenario can be found until the dynamics changes for pump levels of about $p = 1.2$. At this level, an additional higher frequency component emerges at $2.7\,\mathrm{GHz}$, which clearly differs from the relaxation oscillation frequency $\nu_{RO} \approx 0.6\,\mathrm{GHz}$. Further increasing the pump parameter, the low frequency dynamics becomes less and less pronounced, until it vanishes, while the new higher frequency component dominates the dynamics. Nevertheless, the property of the dynamics to evolve cyclically, mediating between stable emission and dynamics, not only persists, but becomes more and more pronounced.

An example of one cyclic scenario for intermediate pump parameters is presented in Figure 4.5. Stable emission is achieved for a pump parameter of $p = 1.33$, which is demonstrated by the corresponding flat rf spectrum depicted in Figure 4.5 a). For slightly increasing the pump level, the intensity starts to oscillate periodically at a frequency of $2.91\,\mathrm{GHz}$. This oscillation becomes more pronounced and slightly shifts to higher frequencies as p is increased to $p = 1.36$. In the rf spectra in Figure 4.5 b), we identify a peak corresponding to the periodic oscillation at its fundamental frequency at $2.92\,\mathrm{GHz}$. Additionally, we find a weakly pronounced second harmonic at $5.84\,\mathrm{GHz}$, while we do not find any dynamics directly related to the relaxation oscillations of

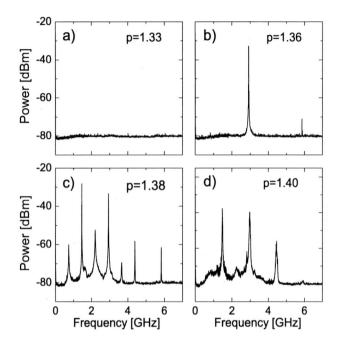

Fig. 4.5: rf spectra of characteristic intensity dynamics of the cyclic scenario for intermediate
pump levels. The dynamics evolves from stable emission in a), to periodic states in
b) and c), to chaotic dynamics d), and then back to stable emission. The feedback
ratio is moderate, $r = 0.16$, and the pump parameters are: a) $p = 1.33$, b) $p = 1.36$,
c) $p = 1.38$, and d) $p = 1.40$.

the solitary laser, which have been determined to be $\nu_{RO} \approx 1.5\,\text{GHz}$. We find that
the fundamental period of the oscillation only slightly depends on p and r, while it is
independent of the choice of resonance condition and, in particular, it is not related
to L_{EC}. This suggests that the oscillations are determined by the dynamical properties of the SL. With further incrementing the pump parameter, we find indications
of a period doubling scenario, which becomes clear from the rf spectrum illustrated
in Figure 4.5 c) for which the pump parameter is $p = 1.38$. For this pump level, we
have already reached a period-4 state of the dynamics, via a period-2 region. In this
case, we find an unusual pronounced first subharmonic of the fundamental frequency
instead of the typical scaling behavior predicted by Feigenbaum [5]. This is probably

due to resonance between the first subharmonic, located at 1.46 GHz, and the relaxation oscillations around 1.5 GHz. We note that for higher pump currents, for which ν_{RO} substantially increases, while the frequency of the first subharmonic only slightly increases, resonance is lost and the peak of the first subharmonic is less pronounced. In the experiments, we can identify period doubling up to the period-8 giving strong experimental evidence for an actual period doubling cascade. Due to the period doubling, we expect chaotic dynamics for higher pump parameters. Indeed, chaotic dynamics can be achieved for $p = 1.4$, which is demonstrated by the rf spectrum given in Figure 4.5 d). The figure discloses that the peaks related to the originally periodic dynamics are significantly broadened. Now, the rf spectrum reveals an enhanced bandwidth of the dynamics of ~ 5 GHz reflecting chaotic dynamics. We note that at the transition form periodic states to chaos the optical spectrum does not give evidence for multimode dynamics. In that sense, the dynamics becomes locally chaotic at the onset of chaos. An additional interesting characteristic of the period doubling route to chaos can be identified from the rf spectrum illustrated in Figure 4.5 d). For increasing the pump parameter and progressing into the chaotic regime, we can identify the inverse cascade. In particular, the residuals of the period-4 peaks are considerably suppressed in Figure 4.5 d). Finally, the cycle is complete, the chaotic dynamics suddenly disappears, and the steady emission state is reached again. This transition between the chaotic and the steady state exhibits hysteresis when decreasing the pump parameter.

The corresponding optical spectra for intermediate pump parameters reveal that with increasing pump parameter the chaotic dynamics also comprises a growing number of LMs extending the optical bandwidth. Therefore, we have substantially increased the pump parameter to verify if the SL system allows for generation of pronounced chaotic intensity dynamics in conjunction with multimode emission comprising high optical bandwidth. In the experiment, we adjust the pump parameter to $p = 3.28$, for which we find pronounced chaotic dynamics. The results are summarized in Figure 4.6. The black line in Figure 4.6 a) depicts the rf spectrum of the dynamics. The continuous spectrum reveals pronounced chaotic intensity dynamics with a bandwidth exceeding 6 GHz. Furthermore, the spectrum does not disclose obvious remnants of periodic frequencies from the period doubling scenario as in the case of lower pump levels in Figure 4.5 d). Now, the transition from stable to chaotic dynamics takes place in a very small interval, which is in contrast to the noticeable period doubling cascade observed for small pump parameters. However, we can identify two specific regions in the rf spectrum. Firstly, there is a broad peak at around 1.3 GHz, which cannot be directly associated with a characteristic system frequency. Secondly, a very broad hump can be identified with a maximum at 3.2 GHz which might be related to the relaxation oscillations, being $\nu_{RO} = 3.06$ GHz for this pump parameter. A 20 ns long segment of the time series is provided in Figure 4.6 b) which illustrates the irregular intensity

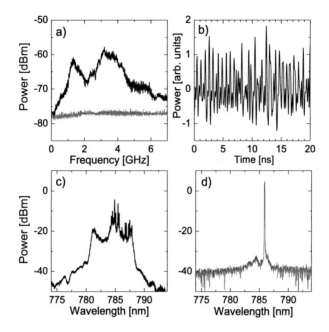

Fig. 4.6: Dynamics for high pump parameter $p = 3.28$ and moderate feedback ratio $r = 0.16$. The rf spectrum of the dynamics, represented by the black line in panel a), reveals broadband chaotic dynamics. Panel b) shows a segment of the corresponding time series, while panel c) demonstrates pronounced multimode emission manifesting itself in the broadband optical spectrum. For comparison, panel d) depicts the optical spectrum for singlemode emission obtained for a slight change of p to $p = 3.25$. The corresponding rf spectrum is represented by the gray line in panel a).

fluctuations on sub-ns time scale. The corresponding optical spectrum is presented in Figure 4.6 c). In contrast to the behavior for small p, the optical spectrum for chaotic dynamics achieved for high p reveals an extraordinarily high spectral bandwidth of about 7 nm. Since the spacing of the LMs amounts to 52 pm, the number of modes involved in the dynamics exceeds 130. An interesting feature of the broadband optical spectrum is its trident-like envelope. We have experimentally verified that this effect does not originate from detuning of the external cavity or dispersion effects in the SL material. This property seems to be a general property for our multimode system. The underlying mechanisms leading to this structure are currently investigated. Furthermore, we note that only for resonant coupling conditions, we are able to achieve such intriguing optical broadband emission. To highlight the dramatic increase in the optical bandwidth, in Figure 4.6 d), we present an optical spectrum for stable emission achieved for small decrementing the pump level by $\Delta p = -0.03$ to $p = 3.25$. This gives also evidence that the cyclic dependence of the dynamics for increasing (and decreasing) pump parameter persists for high pump levels. For completeness, we depict the flat rf spectrum corresponding to stable emission as gray line in Figure 4.6 a). Comparison of both rf spectra reveals that the low frequency part of the rf spectrum for chaotic emission is at the noise floor of detection at $-82\,$dB. This indicates that the average power of the SL light source for optical broadband emission is constant, or more precisely, it exhibits low relative intensity noise with respect to relevant time scales for technical applications.

At first glance, the origin of the observed cyclic character of the scenario for linearly changing p might be surprising. However, a simple explanation can be given by the indirect influence of the pump parameter on the feedback condition. For increasing the pump parameter, the length of the SL cavity $L_{SL,opt}$ slightly elongates because of thermal effects. This causes a small red-shift of the emission wavelength inducing a small decrease of $\Delta\Phi$, which is a cyclic parameter.

Influence of the Feedback Phase

In this subsection, we verify whether the feedback phase is the relevant parameter determining the cyclic nature of the observed scenario. Therefore, we study the effect of small changes of $\Delta\Phi$ on the dynamics. Experimentally, a phase shift can be induced via a small variation of L_{EC} on sub-wavelength scale which can be realized by application of the piezo transducer (PZT), illustrated in Figure 4.1. For proper calibration of the phase parameter $\Delta\Phi$, we have measured the change of L_{EC} in dependence of the voltage supplied to the PZT using a high resolution range meter. We find that a change of $5.5\,$V induces a variation of $\Delta L_{EC} = \lambda/2$ which shifts $\Delta\Phi$ by 2π. We note that for resonant coupling conditions, a proper feedback phase $\Delta\Phi$ is defined, as we have discussed in Subsection 4.2.2. The resonance condition implements the same $\Delta\Phi$ for all LMs under the premise that dispersion in the SL medium is negligible. This is the case for the presented system for which we have measured a maximum dispersion induced deviation from $\Delta\Phi$ of only $\pm 4\,\%$ within the entire optical spectrum.

Figure 4.7 depicts rf spectra of the intensity dynamics for different values of the feedback phase $\Delta\Phi$. In the experiment, we choose operation conditions for fully developed chaotic dynamics. Accordingly, we apply moderate feedback of $r = 0.16$ and adjust for high pumping of $p = 3.28$ for which we measure the relaxation oscillation frequency to

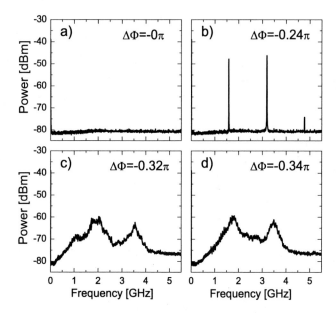

Fig. 4.7: rf spectra of characteristic dynamics of the π-cyclic scenario for decreasing feedback phase $\Delta\Phi$. The phase condition for stable emission, illustrated in panel a), has been assigned to the phase value $\Delta\Phi = 0\,\pi$. Period-2 dynamics for $\Delta\Phi = -0.24\,\pi$ which emerged from period doubling is presented in panel b). Onset of chaos is obtained for $\Delta\Phi = -0.32\,\pi$, depicted in panel c). While fully developed chaotic dynamics is achieved for $\Delta\Phi = -0.34\,\pi$, shown in panel d). Other conditions are $p = 3.3$ and $r = 0.16$.

be $\nu_{RO} = 2.5\,\text{GHz}$. First, we adjust $\Delta\Phi$ for continuous emission by controlling the voltage supplied to the PZT. The corresponding rf spectrum is presented in Figure 4.7 a) and does not reveal indications of dynamics. Since this condition can be easily recognized, we have chosen it as reference and associate it with $\Delta\Phi = 0$. For decreasing the feedback phase from this condition, we find a very similar cyclic scenario as for continuously increasing p. To determine its periodicity, we measure the corresponding phase difference for a cycle of the scenario and find π-periodicity. In the following, we illustrate the emergence of chaotic dynamics within one cycle of the dynamics.

The stable emission state depicted in Figure 4.7 a) is the starting point. From this state, we decrease $\Delta\Phi$ and monitor the influence in the rf spectrum of the dynamics.

For $\Delta\Phi \approx -0.2\,\pi$, we find the onset of periodic dynamics. Then, in agreement with the observations for increasing p, a period doubling scenario evolves with the period-2 state for $\Delta\Phi = -0.24\,\pi$, as illustrated in Fig 4.7 b). For further decreasing $\Delta\Phi$, we find the period-4 state and subsequently a quick transition to chaos around $\Delta\Phi = -0.28\,\pi$. The observed development of the dynamics qualitatively agrees to the period doubling behavior we have discussed in Subsection 4.2.3 for Figure 4.5. However, we note that for this high level of pumping the route to chaos takes place within a very small range of the control parameter between $-0.20\,\pi < \Delta\Phi < -0.28\,\pi$. For further decreasing $\Delta\Phi$, the dynamics evolves into the chaotic regime. This becomes clear from the broad and continuous rf spectrum for $\Delta\Phi = -0.32\,\pi$ depicted in Fig 4.7 c). Nevertheless, even within the chaotic regime the dynamics develops, which is indicated by slight differences in the rf spectra. Fig 4.7 d) presents the rf spectrum of chaotic dynamics for $\Delta\Phi = -0.34\,\pi$, which is close to the sudden transition back to stable emission observed for $\Delta\Phi = -0.36\,\pi$. Approximately 10 indistinguishable cycles of this scenario can be identified for changing the voltage supplied to the PZT. We have also investigated scenarios of the dynamics for different pump parameters and found very good agreement between the dynamics observed for variations of $\Delta\Phi$ and that for changing p. It is worth noting that in all of the experiments we do not find indications of intensity dynamics related to the external cavity round trip frequency $\nu_{EC} = 10.1\,\text{GHz}$. Although the corresponding frequency lies within the detection bandwidth, we find absence of dynamics related to ν_{EC} for different integer and half-integer resonance conditions of up to $M = 5$. This remarkable property might originate from the comparatively slow relaxation oscillations of the SL ($\nu_{RO,max} = 3.3\,\text{GHz}$), which determine the maximum bandwidth of the dynamics.

The sensitive dependence of the dynamics on variations of $\Delta\Phi$ and the cyclic nature of the dynamics suggests similarity to the short cavity regime (SCR) dynamics of non-resonant LK SL systems [86, 87]. However, besides this similarity, we find distinct differences between the conventional SCR dynamics and the dynamics of the resonant short cavities SL system. Firstly, we have identified a period doubling route to chaos instead of the quasi-periodic route being characteristic for SCR dynamics [86,87]. We note that Ryan et al. gave numerical evidence for period doubling for a multimode (5 LMs) non-resonant short cavity LK SL system [115]. However, period doubling has been only identified for weak feedback, which is in contrast to our observations. Secondly, we have experimentally demonstrated π-periodicity instead of the characteristic 2π-periodicity for SCR dynamics. Since this result is quite surprising, we have analyzed the modal structure of the compound cavity system for different resonance conditions and in dependence of $\Delta\Phi$. We find that in the case of half-integer resonance conditions the modal structure for $\Delta\Phi$ is identical to that of $\Delta\Phi' = \Delta\Phi + \pi$, but shifted by $\Delta\nu_{LM}$. Therefore, a phase shift of $\Delta\Phi$ only leads to a change in the dominant group of LMs, as we have discussed in Subsection 4.2.2. For the integer resonance conditions, on the other hand, an identical modal structure is only expected for a phase shift of 2π. In contrast to this, in the experiments, we find similar dynamics for the cycle between $0\pi \leq \Delta\Phi < \pi$ and $\pi \leq \Delta\Phi < 2\pi$. In this case, a closer look at the modal structure of the compound cavity modes for integer resonance conditions $M \geq 2$ indicates similarities for the supported modes of the compound cavity and their neighbors for $\Delta\Phi' = \Delta\Phi + \pi$. This similarity might be origin of this appearing π-periodicity. We note that at the present time, the details about the occurrence of this π-periodic similarity of the dynamics for integer resonance conditions are not fully understood. Insight into this interesting problem could be gained from analysis of appropriate models which also consider the dynamical properties of the SL system. Nevertheless, in both cases the cyclic dependence of the dynamics on the control parameters can be attributed to the cyclic nature of $\Delta\Phi$. Thirdly, the dynamics does not reveal components related to the external cavity round trip frequency ν_{EC} which plays a key-role for conventional SCR dynamics. Finally, the most conspicuous difference between the dynamics of both systems consists in the distinct optical broadband emission, comprising more than 100 LMs, which is unique for the presented system. The emergence of this intriguing property and its relation to the intensity dynamics deserves detailed investigation.

4.2.4 Optical Properties

It is well-known that the occurrence of feedback induced intensity dynamics for conventional LK systems is linked to the emergence of spectral dynamics, which is due to the α-parameter. This results in an enhancement of the linewidth. For this reason, this phenomenon has been termed coherence collapse [118]. However, although this phenomena is well-known, detailed studies on the influence of the dynamics on the coherence properties of ECSLs are rare and the reported studies focus on conventional operation conditions in the long cavity regime [136–138]. However, in analogy to conventional SL systems, we can also expect spectral dynamics in conjunction with intensity dynamics for our resonant cavity system. In the following, we investigate this in detail and study potential interrelations between the intensity dynamics and the spectral properties.

Therefore, we have simultaneously recorded the optical spectra and the intensity dynamics. This allows for investigation of interrelations of both characteristics. Figure 4.8 presents the optical spectra corresponding to the rf spectra of the intensity dynamics of Figure 4.7. The optical spectra have been recorded with a high-resolution interferometric optical spectrum analyzer (IOSA) with a resolution of 4 GHz. This resolution is sufficient to fully resolve the LMs and the external cavity modes with a spacing of

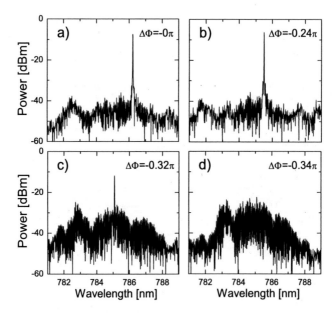

Fig. 4.8: Optical spectra corresponding to the characteristic dynamics scenario presented in Figure 4.7. The parameters are specified in the captions of Figure 4.7.

$\Delta\nu_{LM} = 25.3\,\text{GHz}$ and $\Delta\nu_{EC} = 10.1\,\text{GHz}$. Since the IOSA is based on a Michelson interferometer, it allows in addition for direct measurement of the visibility functions, which are depicted in Figure 4.9. This is beneficial, since this provides direct insight into the coherence properties of the SL system.

In analogy to Figure 4.7 a), we start the analysis of the spectral properties for stable emission at $\Delta\Phi = 0\,\pi$. The resulting optical spectrum is presented in Figure 4.8 a). The spectrum exhibits singlemode emission with a sidemode suppression exceeding 30 dB. Figure 4.9 a) shows the corresponding visibility (VIS) function. The VIS-function is almost constant around maximum visibility and does not give evidence for a fall-off within the accessible range. Hence, the coherence length (L_{coh}) of the emitted light is beyond the resolution of the IOSA. Nevertheless, for stable singlemode emission we can determine L_{coh} by measuring the linewidth with a Fabry-Perot scanning interferometer. The result is presented by the gray line in Figure 4.10. We find a linewidth of $\Delta\nu_{FWHM,fb} = 39\,\text{MHz}$ corresponding to $L_{coh} = 7.8\,\text{m}$, calculated using Eq. 3.13.

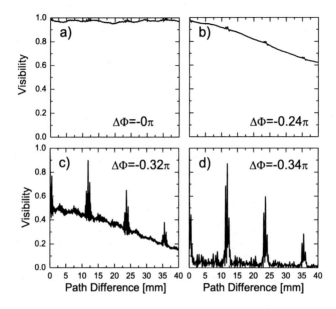

Fig. 4.9: Visibility functions corresponding to the dynamics and the optical spectra depicted in Figure 4.7 and Figure 4.8, respectively.

For comparison, we also measure the linewidth for solitary laser emission for the same parameters, but in the absence of feedback. The experimental result is depicted as black line in Figure 4.10. The spectrum reveals a linewidth of $\Delta\nu_{FWHM,sol} = 52\,\mathrm{MHz}$ from which we deduce a coherence length of $L_{coh,sol} = 5.8\,\mathrm{m}$. This means that the coherence length of the SL system for stable emission can be extended by application of feedback from a resonant cavity.

However, the spectral properties appreciably alter for tuning $\Delta\Phi$ away from the stable emission state when entering the regime of dynamics. Figure 4.8 b) presents the optical spectrum for the period-2 state for $\Delta\Phi = -0.24\,\pi$, slightly beyond the onset of dynamics. Interestingly, the optical spectrum remains singlemode, but with slightly broadened linewidth. This results in considerable reduction of the coherence length, as it can be seen in Figure 4.9 b). In this case, the coherence length can be extrapolated to $L_{coh} \approx 0.7\,\mathrm{m}$. Further decreasing $\Delta\Phi$ within the regime of periodic intensity dynamics continuously enhances the linewidth and reduces the coherence length. We

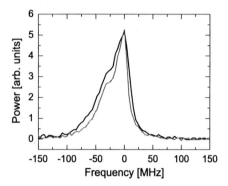

Fig. 4.10: Comparison of the optical linewidth for stable emission for $p = 2.9$. Solitary laser emission without feedback is presented in black, while stable emission for $r = 0.16$ and adjusted feedback phase is presented in gray. The linewidth for solitary laser emission amounts to $\Delta\nu_{FWHM,sol} = 52\,\text{MHz}$ and is reduced by application of feedback to $\Delta\nu_{FWHM,fb} = 39\,\text{MHz}$.

note that for periodic dynamics, we find hints for very weakly pronounced external cavity side modes near the central LM. At the onset of chaos for $\Delta\Phi = -0.32\,\pi$, we find reduced intensity for the dominant LM, but also numerous LM sidemodes that are suddenly involved in the dynamics. This feature is illustrated in Figure 4.8 c) in which we identify that the intensity of the LMs in the range between 781.8 nm and 788.2 nm has considerably increased. The spectrum is very stable, while its noisy appearance is again due to the high resolution of the IOSA revealing more than 120 LMs participating in the emission. This coupling of the LMs drastically affects the coherence properties, which is depicted in Figure 4.9 c). The VIS-function discloses peaks at multiples of 11.9 mm. This path difference equals twice the optical length of $L_{SL,opt}$. These peaks have also been reported by Daza [137] for conventional multimode SL with optical feedback, but the authors did not provide an explanation of the occurrence of these peaks. The envelope of the peaks describe a curve with similar shape as the one presented in Figure 4.9 b), but with faster fall-off. Additionally, Figure 4.9 c) reveals a drastic decay of the VIS within 550μm down to $VIS \approx 0.5$. Between the peaks, the VIS continuously decreases for increasing path difference. This sudden reduction of the visibility occurs as soon as the LMs couple inflating the spectral bandwidth. Further increasing the feedback phase to $\Delta\Phi = -0.34\,\pi$, the intensity dynamics in Figure 4.7 d) reveals fully developed chaos. The corresponding optical spectrum is depicted in Figure 4.8 d).

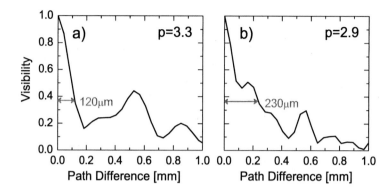

Fig. 4.11: Influence of the pump parameter on the coherence properties in terms of the visibility function for moderate feedback ratio $r = 0.16$. Panel a) depicts a zoom into Fig. 4.9 d) for $p = 3.3$. Panel b) shows a zoom into the corresponding VIS-function measured for $p = 2.9$.

The spectrum reveals pronounced multimode emission comprising about 100 LMs in a spectral range of ~5 nm, which suggests further reduction of the coherence. Indeed, the visibility function in Figure 4.9 d) exhibits fast fall-off of the VIS below 1/e within only 120 μm. This is illustrated by a zoom into the corresponding visibility function, which is presented in Figure 4.11 a). The figure reveals the fast decay of the VIS within 120 μm. However, we also notice a peak of recurrent VIS at a path difference of about 530 μm. This length does not coincide with any optical length of the devices of the SL system. Therefore, we exclude external feedback effects as being the origin for this peak. Additionally, we exclude dynamical effects, since the corresponding frequency amounts to 505 GHz which is at least one order of magnitude beyond the bandwidth of the dynamics. The occurrence of this VIS peak probably originates from a defect within the cavity of the SL. Such a defect can act as an internal reflector forming a sub-cavity within the SL cavity. We observe a slight dependency of the corresponding VIS peak on the alignment of the external cavity which supports this consideration. By doing this, we can suppress this peak below values of 1/e. This assumption has been verified by similar measurements with other lasers which did not exhibit this additional feature in the VIS-function. Hence, this disturbing recurrent VIS in the present systems could be avoided in practical systems. Therefore, Figure 4.11 a) demonstrates that it is possible to realize short coherence lengths of approximately 100 μm. Additionally, we find that the envelope of the peaks depicted in Figure 4.9 d) also gives rise to a slightly faster fall-off, if compared to that of Figure 4.9 c). We note that these peaks, or remnants, of VIS can be reduced by increasing the dynamical bandwidth of the SL system. Therefore, one can increase the pump parameter to obtain faster relaxation oscillations. Another possibility is to exchange the laser with another SL offering faster relaxation oscillations and, hence, higher dynamical bandwidth. The other way round, decreasing the bandwidth of the dynamics results in higher coherence of the emitted light. This induces a slower fall-off of the VIS-function, as it can be deduced from the VIS-function presented in Figure 4.11 b), which we obtain for slightly reduced pump parameter of $p = 2.9$. For this condition, we find a slower fall-off of the VIS to values below $1/e$ of $230\,\mu$m. Furthermore, the envelope of the VIS remnants related to the LMs also discloses slower fall-off. Since these remnants are related to the spacing of the LMs, this property of the VIS-function already indicates changes in the correlations between adjacent LMs, which can be influenced by modification of the dynamical bandwidth of the system and the coupling conditions between the LMs. We will study the interactions between the LMs in the next subsection.

The presented results are interesting from both the fundamental NLD and the application-oriented point of view. The experiments have revealed a strong interrelation between the intensity dynamics and the spectral dynamics. Interestingly, the dynamics emerges in a spectrally well-confined region in the vicinity of the lasing LM. In this regime, the dynamics only enhances the linewidth of the central LM which, in turn, reduces the coherence length. At the onset of chaos, the spectral dynamics in the vicinity of the central LM suddenly starts to inflate because of coupling of numerous LMs. For fully developed chaos, the LMs strongly couple and multiple LMs participate in the dynamics giving rise to rather global dynamical behavior. The onset of this characteristic multimode emission is fascinating since it might reveal general properties of pronounced multimode dynamics which have also been reported for other laser systems, such as for fiber lasers [103]. Intuitively, the pronounced multimode emission can result from spectral overlap of the dynamics of adjacent LMs. Since for resonant feedback $\Delta\Phi$ is similar for all the LMs, the relative spectral position between the external cavity modes and the LMs is similar for all LMs. This means that once the central LM couples to its neighboring LMs, these LMs also couple to their neighbors and so forth. However, further investigations are necessary to fully understand the underlying mechanisms. Such understanding is also required for optimization of the emission qualities with respect to technical applications.

We have demonstrated that the coherence properties of this systems can be efficiently controlled and varied in a wide range between approximately $100\,\mu$m and $10\,$m by application of NLD. The resonant short cavities SL system offers excellent qualities for implementation of chaotic light detection and ranging (CLIDAR) [71] and coherence tomography [146], since it allows for rapidly decaying VIS within the range of $\sim 100\,\mu$m. Although the applied principle for reduction of the coherence length is based on chaotic emission dynamics, the average output power of the light source is constant on the technically relevant time scales. This can be deduced from Figure 4.6 a), which shows that the low frequency part of the dynamics is at the noise floor of detection. This means that the relative intensity noise of the light source is low on the relevant timescale of the proposed applications. Furthermore, the light source exhibits the good beam properties of a laser and offers a maximum output power of 100 mW. We note that there are also recurrent regions of higher visibility at multiples of $2L_{SL,opt}$. For high-resolution ranging, the fast fall-off of the visibility is relevant with a corresponding coherence length of $\sim 100\,\mu$m. Nevertheless, other applications might exist for which the visibility peaks may become of importance. From the NLD point of view, in turn, the recurrent peaks of the visibility are very fascinating. The fact that the spacing of the peaks $\Delta L_{peak} = 2L_{Sl,opt}$ is related to the spacing of the LMs $\Delta \nu_{LM} = c/\Delta L_{peak}$ already suggests that the dynamics of the LMs is correlated. This observation raises the question of possible interactions between the LMs and their role for the occurrence of pronounced multimode emission.

4.2.5 Interactions of the Longitudinal Modes

In this subsection, we study interactions between the LMs for the optical broadband dynamics. Therefore, we perform spectrally resolved measurements, similar to the techniques presented in references [93, 152–154]. In the experiment, we simultaneously acquire the total intensity dynamics, comprising all the LMs, and spectrally filtered dynamics, comprising the dynamics of only 7 LMs. Then, we compare both dynamics and repeat the procedure for different filter positions. From this, we gain insight into the role of the particular spectral components to the total intensity dynamics. In the experiment, we realize spectral filtering using the grating monochromator (MO) illustrated in Figure 4.1. We chose similar conditions as for the previous experiments: moderate feedback of $r = 0.18$ and moderate, but slightly reduced, pumping of $p = 2.52$. Additionally, we adjust $\Delta\Phi$ for maximum optical bandwidth.

Figure 4.12 summarizes the results of the experiment. In the figure, the total dynamics is presented in gray, while the spectrally filtered dynamics is represented in black. To provide a complementary overview over the dynamics, we present the optical spectra, the rf spectra, and 15 ns long normalized time series of the intensity dynamics for measurements with three different filter positions. Each of the horizontal rows in Figures 4.12 represents the results for one of the three filter positions which are: 781.7 nm, 784.8 nm, and 787.3 nm. The optical spectra are depicted in the Figures 4.12 a), b)

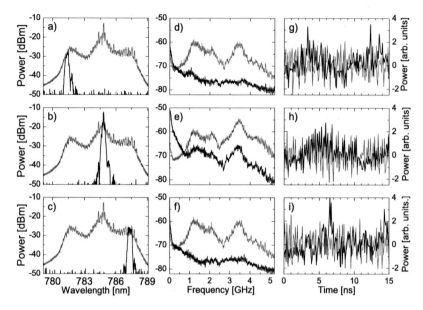

Fig. 4.12: Comparison of total and spectrally filtered dynamics for optical broadband emission. The three different center frequencies of the filter correspond to 781.7 nm in a), 784.8 nm in b) and 787.3 nm in c). The total dynamics is represented in gray and the filtered dynamics in black. The optical spectra are depicted in a), b) and c), while the corresponding rf spectra and the normalized time series are presented in Figures d)-f) and g)-i), respectively. Other parameters are: $p = 2.52$ and $r = 0.18$.

and c) and exhibit an optical bandwidth of approximately 7 nm for the total dynamics. The 3 dB bandwidth of the filtered dynamics has been determined to be 350 pm. Accordingly, the spectrally filtered dynamics reflects the dynamics of 7 LMs.

We start with the comparison of the total dynamics and the filtered dynamics in vicinity of the center wavelength of the optical spectrum at 784.8 nm (b,e,h). The rf spectrum of the filtered dynamics in Figure 4.12 e) reveals conspicuous dynamics for frequencies below 1 GHz which is lacking in the rf spectrum of the total dynamics. Besides this, both rf spectra show good qualitative agreement for the fast components of the dynamics between 2 GHz and 5 GHz. We note that the different power of the full and the spectrally filtered dynamics about 10 dB is due to the loss of power because of spectral filtering, as it can be seen from the optical spectra in Figure 4.12. These two characteristics can also be identified by comparison of the corresponding time series which are presented in Figure 4.12 h). On the one hand, the time series for the filtered dynamics exhibits slow intensity fluctuations on a time scale of several ns. Such slow fluctuations are absent in the time series of the total dynamics, giving evidence for anti-phase dynamics of the LMs [155]. This phenomenon is well-known for less pronounced multimode SL systems [93, 100, 156, 157]. In many of these systems, competition of the LMs for the common gain has been identified as the origin of anti-phase dynamics, since it induces considerable coupling between the LMs [127]. In contrast to the anti-phase dynamics at low frequencies, we find in-phase dynamics in the time series for the fast pulsations on sub-ns time scale. Such in-phase dynamics has also been reported for weakly pronounced multimode SL systems. For these systems, it has turned out that in-phase dynamics was related to the relaxation oscillations [93]. However, for the chosen operation conditions the relaxation oscillation frequency amounts to $\nu_{RO} = 2.6\,$GHz, which substantially differs from the dominant peak at 3.5 GHz in the rf spectra. This different peak frequency can arise from feedback and multimode effects which influence the in-phase dynamics in this frequency range. Further insight into the interactions of the LMs is required for clarification of this problem. Therefore, we study the dynamical properties of spectral regions which are apart from the center of the optical spectrum. In the corresponding experiments, first, we decrease the center frequency of the filtered dynamics to 781.7 nm. Then, we increase the center frequency of the filter to 787.3 nm. The results are presented in Figure 4.12 in (a,d,g) and in (c, f, i), respectively. The rf spectra are presented in Figures 4.12 d) and f). Both spectra of the filtered dynamics reveal more pronounced low frequency dynamics than the total intensity dynamics[1]. This agrees with our previous finding for the filtered dynamics near the center of the optical spectrum presented in Figure 4.12 e). In contrast to the dynamics near the center, the dynamics at the flanks of the optical spectrum evidences substantially less pronounced high frequency dynamics. This property is also reflected by the corresponding normalized time series, which are depicted in the Figures 4.12 g) and i). In both cases the time series of the filtered dynamics exhibit low frequency intensity dynamics. In contrast to the intensity dropouts on slow time scales which can be recognized in Figure 4.12 h), the slow dynamics for the spectrally distant regions consists of short intervals of increased power. The fast intensity pulsations, on the other hand, are considerably less pronounced. Furthermore, we do not find indications of distinct in-phase dynamics in time series of the spectrally distant regions.

The results indicate that mode competition is also a relevant mechanism for the reso-

---

[1]The pronounced low frequency dynamics of the spectrally filtered dynamics becomes apparent when the power spectra of the filtered dynamics are normalized to the power of the total dynamics, considering the loss of power due to spectral filtering.

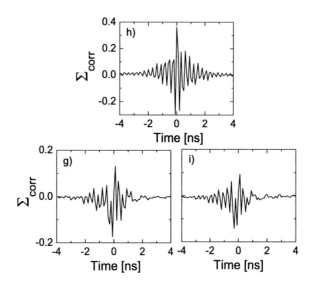

Fig. 4.13: Correlation functions between the total intensity and the spectrally filtered time
series of the measurement presented in Figures 4.12 g), h), and i). The center
frequencies of the filter are 781.7 nm in g), 784.8 nm in h), and 787.3 nm in i).

nant short cavities SL system. Even more, we find evidence for considerable interactions
of the LMs within the complete spectral range. In contrast, the experiment demonstrates that the dynamics on the fast time scales beyond 1 GHz originates more from
emission near the center of the optical spectrum. These results immediately raise the
question, whether the dynamics is inherently broadband for all LMs disregarding the
spectral position, or if the dynamics in the center drives the dynamics in spectrally
distant regions. In terms of information theory based NLD, this question is directly
related to the question about the generation and the flow of information. The answer
to this problem provides fundamental insight into the modal interplay in pronounced
multimode systems [158].

First access to this problem can be obtained from correlation analysis of the recorded
time series. Figures 4.13 g), h), and i) present the calculated cross-correlation functions
(Σ_{corr}) between the time series of the total intensity dynamics and that of the spectrally
filtered intensity dynamics shown in Figures 4.12 g), h), and i). In this representation,
high correlation at negative times means that the spectrally filtered dynamics lags with

respect to the total intensity dynamics. The cross-correlation for the total intensity dynamics and the spectrally filtered dynamics of the central region is presented in Figure 4.13 h). Σ_{corr} exhibits two features. Firstly, its substructure reveals several peaks which can be assigned to the fast dynamics with a dominant component around 3.5 GHz, as it can be seen in Figure 4.12 g). Secondly, we identify a far reaching envelope, which also gives rise to correlation between slower dynamics components. In this case, we find an almost symmetric envelope of the cross-correlation function and maximum correlation at zero lag. These properties indicate that the dynamics in the central region is simultaneously generated on both the slow and the fast time scales, if compared to the total intensity dynamics.

The correlation behavior changes for the spectrally distant regions is depicted in Figures 4.12 g) and i). In similarity to the central spectral region, for both cases, we find coinciding maximum cross-correlation at zero lag. This shows that the fast dynamics is also generated simultaneously with the fast dynamics generated near the central spectral region. In addition, the envelope of the cross-correlation function, on the other hand, reveals an asymmetric envelope with long-range, dominantly negative correlations for negative delays. This property can be attributed to comparably slow anti-phase dynamics which lags with respect to the total intensity dynamics. Further experiments are required to gain deeper insight into the fundamental properties of the multimode interactions. In particular, cross-correlation measurements between fully resolved dynamics of spectrally distant regions can provide important information about the mechanisms underlying the pronounce multimode dynamics. These measurements are experimentally demanding and are subject to current investigations. Additionally, analysis of suitable models is a desirable, complementary approach to achieve this goal. Nevertheless, a suitable simple model, comparable to LK based modeling, which is capable to account for the relevant resonant coupling condition is currently not available.

4.3 Concluding Summary

We have presented detailed studies on the emission properties of a resonant short cavities SL system with comparable lengths of the external cavity and the semiconductor laser cavity. We have demonstrated that resonant coupling between both cavities efficiently enhances coupling between the LMs allowing for pronounced multimode emission. For resonant coupling, we have identified the feedback phase $\Delta\Phi$ as being a major control parameter determining the dynamics of the SL system. For continuously decreasing $\Delta\Phi$, the intensity dynamics evolves in a π-cyclic scenario from stable emission to periodic states, to chaos, and again back to stable emission. Analysis of the resonant short cavities dynamics revealed conspicuous differences if compared to the well-known short cavity regime dynamics exhibited by non-resonant SL systems [86,87]. A remarkable property of the presented system consists in the possibility of generating broadband chaotic intensity dynamics in conjunction with distinct multimode dynamics comprising an optical bandwidth of $\sim 7\,\text{nm}$. For this state, more than 130 LMs participate in the dynamics. Detailed analysis of the spectral properties of the dynamics revealed strong interrelations between the intensity dynamics and the spectral emission properties of the system. We have demonstrated that this property can be utilized for controlled adjustment of the coherence properties of the light, which offers an accessible range of coherence length between $\sim 100\,\mu\text{m}$ and $\sim 10\,\text{m}$. In that sense, we have demonstrated that NLD can be beneficially harnessed for realization of light sources with customized emission properties. With respect to technical applications and from the fundamental point of view, understanding of the underlying mechanisms leading to the emergence of such extraordinarily pronounced multimode dynamics is desired. Consequently, we have performed spectrally resolved measurements of the dynamics that provided first insight into the complex dynamics. The results revealed considerable interactions between the LMs manifesting themselves in anti-phase dynamics on time scales slower than nanoseconds. This result has been verified by detailed cross-correlation analysis, which has also given evidence that the fast components of the dynamics are simultaneously generated within the total spectral range.

From the NLD point of view, identification of the underlying mechanisms leading to such pronounced multimode dynamics is desired, since it might reveal general properties of complex multimode system. Complete understanding of the dynamics requires further experiments such as fully spectrally resolved measurements allowing for analysis of inter-LM correlations. Such experiments are challenging and impose high demands on the measurement technology. Therefore, practical models are desired which can provide complementary insight. However, suitable models which can account for the resonant coupling condition giving rise to pronounce multimode dynamics are currently not available.

We point out that the demonstrated coherence properties of the system indeed are highly interesting for novel technical and medical applications. Therefore, SLs with delayed optical feedback provide an interesting alternative to existing incoherent light sources. The specifications of the realized light source directly offer implementation in applications such as chaotic light detection and ranging applications [71], and coherence tomography [146]. Furthermore, future implementations of such SL light sources in consumer electronics are imaginable, e.g., for illumination of high-resolution displays or laser projectors for which speckle-reduction is of importance. Finally, we note that the performance of the system offers potential for improvement. In addition to an optimization of the well-controllable operation parameters, laser device characteristic parameters can be adapted to the requirements of particular applications. In particular, application of recently available quantum dot lasers with novel gain structures that offer distinctly higher gain bandwidth suggest promising results, if suitable α-parameters can also be realized. We note, that latest measurements reported α-parameters for these SLs in a range between $0 < \alpha < 25$, supporting feasibility of this promising approach.

Chapter 5

Encrypted Communications Utilizing Chaotic Light

We have seen that semiconductor lasers with delayed optical feedback are well-suited for the controlled generation of high-dimensional broadband chaotic emission dynamics. It has turned out that the emission properties of ECSL systems can be well controlled by utilizing NLD. This allows for manipulation and optimization of the dynamical properties according to the demands of specific applications. In this chapter, we study the potential of a fascinating application of chaotically emitting ECSL systems in which the properties of the chaotic light are harnessed for realization of encrypted communication systems. The functionality of such so-called *Chaos Communication Systems* is based on two essential properties of chaotic systems. Firstly, the intrinsic irregularity of chaotic dynamics can be utilized for encryption of messages, while the chaotically emitted light of ECSLs can simultaneously serve as message carrier signal. Secondly, similar chaotic systems can synchronize their dynamics when the systems are coupled [159, 160]. This somewhat counterintuitive behavior of coupled chaotic systems is known as *Chaos Synchronization*. Chaos synchronization represents the fundamental mechanism on which the principle of chaos communication is based. Functionality of a very promising class of chaos communication schemes is based on so-called *Chaos-Pass-Filter* properties. In these schemes, a receiver system, capable of synchronizing to the dynamics of the transmitter system, replicates the carrier signal, while it does not replicate the message which has been mixed into the carrier. This property represents the crux of very promising approaches to realize functional chaos communication systems, since it allows for discrimination of the message and the carrier signal at the receiver and, hence, decryption of the message.

In the following, we comprehensively analyze the properties of the necessary constituents to realize a functional chaos communication system. Chapter 5 is organized as follows: After a brief introduction into the principles of chaos communications, we discuss the requirements on the chaotic carrier signal generated by the transmitter system. From these requirements, we show how the operation conditions of the ECSL system can be optimized to generate appropriate high-dimensional broadband chaotic

135

carrier signals. Then, we study two different receiver configurations which we refer to as the *Open Loop* and the *Closed Loop* scheme. We demonstrate first successful synchronization of the receiver dynamics to such optimized complex chaotic transmitter dynamics for both schemes and analyze their synchronization properties with respect to the realization of a practicable chaos communication system. Our results reveal excellent, synchronization properties for both configurations offering alternative message encryption and decryption techniques, which we study in detail. Based on these results, we investigate the feasibility of different encryption/decryption methods under consideration of performance and security aspects. These results obtained results allow for first realization of a practicable chaos communication system which allows for security levels and fast transmission rates. This will be demonstrated by successful encryption, transmission and recovery of a binary signal. The successful realization of an optical cryptosystem emphasizes the functionality of the chaos communication principle, furthermore underlining the high potential of applied NLD for modern technology. Additionally, our results provide insight into the phenomenon of chaos synchronization which is important for diverse nonlinear dynamical systems occurring in nature.

5.1 Principles of Chaos Communication

Over the last 15 years, there has been a rapidly growing interest in chaos synchronization phenomena. Initially, research was mainly motivated by the intention to understand the mechanisms of the counterintuitive possibility to synchronize the dynamics of independently chaotic dynamical systems when introducing a coupling mechanism [160, 161]. Synchronization of chaotic systems has turned out to exhibit a much richer phenomenology than synchronization of periodic systems. A variety of fascinating synchronization phenomena [162] has been discovered in various areas of science, e.g., in physics [163, 164], chemistry [165], biology [166] and physiology [167, 168] [1]. Their occurrence has attracted much interest and boosted research in this very active branch of nonlinear science. Beyond the interest from the fundamental-science point of view, chaos synchronization phenomena open perspective for novel application concepts [169, 170].

Chaos communication is currently one of the most promising applications utilizing chaos synchronization phenomena. The basic idea of using chaotic signals as message-carriers for data transmission was introduced by Pecora and Carol in 1990 [171]. Since then various chaos communication systems based on different functional principles have been proposed. One of the most promising scheme for realization of chaos communi-

---

[1] An overview over synchronization in numerous diverse systems is provided in [31, 32].

cation systems can be considered as a generalization of conventional communication systems in which a message is modulated onto a periodic transmitter carrier signal, termed carrier in the following, and sent to a receiver. In conventional communication systems, the receiver is tuned to the periodic transmitter signal, so that it is possible to discriminate the message from the carrier signal, which is usually done electronically. This means that detailed knowledge about the carrier signal is essential for successful message recovery. Some advanced spread-spectrum communication techniques apply carriers with broadband frequency spectra to reduce the liability of the transmission-system towards spectral disturbances. In compliance with this, carriers generated by chaotic transmitters are also broadband being an intrinsic property of chaotic dynamics. Therefore, they represent potential carrier signals for reliable communication systems. Furthermore, in Chapter 3, we have demonstrated that chaotic waveforms with maximum frequencies of the order of 10 GHz can be realized being potential carriers for multi-Gbit/s rate data transmission. In similarity to conventional communication systems, in chaos communication systems the message extraction process is also based on synchronization phenomena. In fact, it is by far more challenging to achieve synchronization of chaotic systems, but the chaos synchronization offers also beneficial properties allowing for realization of encrypted communication systems. In these systems, the receiver needs to synchronize to the carrier signal of the transmitter to extract the message. This method offers major advantages if compared to conventional communication schemes. One advantage is that the message can be nonlinearly mixed within the carrier which provides a dynamical key inherent to the transmission of the message. The encryption properties of chaos communication systems strongly depend on the way the message is introduced into the nonlinear transmitter system. Different encryption methods have already been proposed which are known under as Chaos Masking, Chaos Modulation Keying, Chaos Shift Keying, and ON/OFF Shift Keying [172]. Later on in this chapter, we will explain these techniques and discuss their properties. Finally, chaos synchronization is required for successful message extraction, since only synchronized receivers can replicate the carrier signal so that discrimination between the message and the chaotic carrier is possible [80, 171]. This is a major advantage of chaos communication systems against conventional communication schemes, because only a receiver very similar to the transmitter, in parameters and operation conditions, can synchronize to the chaotic carrier. Hence, the knowledge of the receiver structure and its operation parameters provides a dynamical key which is required for message extraction. This means that chaos communication systems represent cryptosystems which allow for message encryption on the physical layer, which does not require time consuming computational efforts as it is required for software encryption. We note that, chaos communication is fully compatible with software encryption. Even more, chaos communications represents a complementary and fully compatible approach to absolutely secure quantum cryptography. The advantage of quantum cryptography is that it can allows for absolutely secure transmission of information. Unfortunately, the transmission rates which can be achieved at the present time do not compete with telecommunication standards. Therefore, currently the investigations in this active field of science aim for application in secure exchange of keys which are used for subsequent (software) encryption of messages. In this context, chaos communication and quantum cryptography offer potential for promising joint approaches. In such an approach quantum cryptography could facilitate secure exchange of the key(-parameters) which are required to synchronize the chaotic sub-systems of a chaos communication system. The chaos communication system can then be used for transmission of encrypted messages at high-bit rates. In that sense, such a joint approach for realization of high-performance system for encrypted data communication can benefit from the advantages of both, the security of quantum cryptography and the fast encryption and transmission rates which can be realized with chaos communication, as we will see in the following.

In recent years, various approaches have been followed for the realization of optical communication systems [79, 80, 170, 171, 173–176]. To impede message recovery by eavesdroppers using advanced time series analysis methods, scientific interest has soon focused on chaotic systems exhibiting high-dimensional dynamics. In this context, it has promptly been realized that laser-based feedback-systems are well-suited to exhibit such dynamics. The first optical chaos communication system was realized by Van-Wiggeren and Roy in 1999 using fiber lasers [170, 177]. However, it did not take long until Geddes and Short [178] presented results which demonstrated that it is possible to crack the code of the proposed encryption scheme. Because of these results, it has been realized that the dynamical properties of chaotic carrier signals for functional chaos communication systems need to fulfill additional specific requirements if high-level of security is desired. Since then, a growing interest in the particular properties of the chaotic carriers has emerged, since it has turned out that they crucially affect functionality, applicability and security of chaos cryptosystems. For applications in high-speed data transmission-systems, for instance, where maximum bit rates of multiple-Gbit/s and high reliability are desired, the maximum frequency and the bandwidth of the carrier should be as high as possible. Other characteristics of the carrier are relevant for the system's security, e.g., dimensionality, nonlinearity and auto-correlation properties [178–181]. Hence, for a successful implementation of a chaos cryptosystem all the performance aspects comprising carrier, encryption method, synchronization properties and decryption techniques have to be considered, which has not been done so far.

As it can be suggested from the results presented in Chapter 3, semiconductor laser

(SL) systems offer particularly high potential for implementation in modern chaos cryptosystems. Therefore, we will follow this promising approach, since SL systems allow for generation of high-dimensional broadband chaotic carrier signals. In addition to this, SLs are standard telecommunication devices. Thus, many of the optical and the electro-optical components which are required for realization of a chaos cryptosystems are available of-the-shelf. Furthermore, utilization of existing fiber links seems to be accomplishable and of high value from an economical point of view. These advantages of SL systems have attracted much interest for realization of a chaos communication system, but also for fundamental scientific reasons in the field of NLD. First evidence of chaos synchronization in SLs was given numerically in 1990 when Winful et al. studied laterally coupled SLs [161]. In 1996 Mirasso et al. and Annovazzi-Lodi et al. gave first numerical evidence for the possibility of chaos communication with spatially distant SL systems [78, 79]. Since then, synchronization of the intensity dynamics of many different SL systems has been studied [80, 182–184], but also synchronization of other dynamics such as synchronization of wavelength chaos [185] has been reported.

So far, several chaos cryptosystems utilizing SLs subject to delayed feedback have been proposed, comprising different configurations and diverse encryption/decryption methods [78,174,175,186]. Some systems have been experimentally implemented and a proof of principle as cyroptosystems has been demonstrated [187–190]. Nonetheless, detailed analysis of the influence of the dynamical properties, the encryption/decryption method, and the synchronization properties on the overall performance of the chaos communication systems has been lacking. However, it has turned out that optimal performance of a functional chaos communication system might requires a tradeoff between practicability and security. To allow for such optimization, it is essential to identify the key-features which determine the performance aspects and to investigate their dependence of the relevant system parameters.

In the following, we will address this problem and present a comprehensive analysis and characterization of the relevant system properties that need to be considered for designing a functional system. Therefore, we first define requirements on a practicable chaos communication systems. These requirements concern the characteristics of the carrier signal, the synchronization properties, and the qualities of the encryption/decryption method. Subsequently, we study whether the requirements can be experimentally accomplished. Our results lead to different favored configurations which demonstrate that the performance properties of functional chaos communication systems can be designed according to the desired level of security and transmission rate.

Physical Properties	
Performance	broad bandwidth
	well-controllable parameters
	high carrier modulation index
Privacy	broad bandwidth
	flat rf spectrum
	rapidly decreasing correlation
	high-dimensional attractor
	high information entropy
	suitable encryption technique
	selective on parameter tuning

Table 5.1: Requirements on carrier signals for chaos communication systems.

Requirements on Carrier Signals

The generation of tailored chaotic carrier signals, which are suitable for message encryption with GHz bandwidth, is one of the fundamental challenges for realization of applicable chaos cryptosystems. Reason for this is that the carrier signal properties strongly influence the performance of the communication systems. Thus, special care has to be taken when designing a carrier signal generator to meet the desired characteristics of the chaos communication system. For that reason, we present and discuss requirements on carrier signals for chaos communications. These requirements will serve as guideline for realization and optimization of the carrier signals which can be generated by ECSLs.

Table 5.1 summarizes central requirements on the physical properties of chaotic carriers signals. These requirements can be subdivided into two categories concerning performance and privacy aspects. From the performance point of view, sufficient bandwidth, high maximum carrier frequency and good controllability of the key parameters of the dynamics are desirable. With respect to privacy, additional requirements need to be fulfilled. The most important ones consist in high information entropy, high-dimensionality, and rapidly decreasing auto-correlations of the dynamics. In this way, it is possible to impede potential eavesdropping that is based on reconstruction of the strange attractor of the dynamics by using phase space embedding techniques [178, 179, 181]. Furthermore, the carrier signal needs to allow for realization of practical encryption methods to guarantee high levels of security. Additionally, the technical implementation should be practicable to allow for realization of a robust and functional system. Our aim is to generate chaotic carrier signals which meet the requirements presented in Table 5.1 as good as possible.

We try to achieve this goal by using chaotically emitting ECSLs, since these systems have already proven their suitability for the generation of high-dimensional broadband chaotic dynamics. Another advantage of ECSLs is that they are experimentally well-controllable so that their dynamical properties can be fine-tuned. This is an essential property for chaos synchronization experiments for which synchronization can only be achieved for well-matched dynamical systems, implying that the strange attractors of the transmitter and the receiver system have similar geometrical and dynamical characteristics. Since chaos synchronization is the fundamental mechanism on which the principles of chaos communication are based, there are also specific requirements on the synchronization properties of the system.

Requirements on Synchronization Properties

Over the last 15 years, chaos synchronization phenomena have attracted much interest in the scientific community because of their fundamental importance for the understanding of nature and for functionality of many applications. Since Christian Huygens discovered synchronization in 1665, while observing two pendulum clocks hanging side by side on a wooden wall, synchronization phenomena of coupled oscillators have been intensively studied. Today, the synchronization phenomena of coupled chaotic oscillators that occur in a diversity of systems in nature are still not fully understood. Because of their relevance, investigation of these phenomena is the focus of ongoing research. Chaotic oscillators can exhibit different synchronization phenomena including different types of synchronization. The types of synchronization reflect different qualitative properties of synchronization. The most straightforward type thereof is referred to as complete synchronization (CS) in which the trajectories of the dynamics of the synchronized systems are identical in phase space [159, 160, 162, 191]. For this type of synchronization, the frequency and the phase of the oscillators are completely locked. In a weaker form of synchronization either the frequency or the phase of the oscillators is locked. These types of synchronization are known as frequency synchronization (FS) [159] and phase synchronization (PS) [192]. In addition to that, chaotic oscillators can also exhibit synchronization types in which the phase and the amplitude of the dynamics of the oscillators are related by some well-defined interdependence. Such a state of synchronization is referred to as generalized synchronization (GS) [193][2]. In the weakest form of synchronization the amplitude of chaotic oscillators is not correlated, but the phase of one of the oscillators gives rise to a weak dependence of the dynamics of the other oscillator. In this case, the phase of the driven oscillator follows that of the driving oscillator, but with delay. Therefore, this type of synchronization has been named lag-synchronization (LS) [162]. In this framework perfect and imperfect synchronization can be distinguished and the quality of synchronization can be categorized. Nevertheless, we also point out that the transitions between these states of synchronization are usually continuous.

It is not surprising that the different of types of synchronization gives rise to various characteristic synchronization phenomena, all of which are highly interesting from the fundamental NLD point of view. However, not all of the synchronization qualities are suitable for our approach for realization of functional chaos communication systems. We are interested in a robust form of synchronization that allows for best possible reconstruction of the transmitted chaotic carrier signal in the receiver, so that the message can be recovered. Fulfillment of this purpose requires specific synchronization properties:

- The synchronization needs to be robust against small perturbations. This implies the existence of a stable synchronization manifold of the system, guaranteeing reliable operation of the communication system.

- The functional relation between the transmitter and the synchronized receiver needs to be well-defined. This requirement allows for reconstruction of the carrier signal from the response of the synchronized receiver. A simple functional interdependence between the dynamics of both system facilitates decryption of the message.

- The time it takes to gain (and lose) synchronization should be as short as possible. In the case of temporal loss of the synchronization, the synchronization should be recovered as fast as possible to prevent significant loss of the message. The transient times for loss and regain of synchronization need to be minimized, since they limit the bandwidth of the message transmission.

These basic requirements on the synchronization properties not only facilitate reliable operation of specific chaos communication system, but even more they open the perspective for realization of fundamentally different chaos communication schemes.

---

[2]A detailed description of the difference between perfect and generalized synchronization is provided in Appendix A.2.

Essentially, one can say that good synchronization qualities are desired for realization of a practical robust chaos communication system. Therefore, perfect synchronization or good generalized synchronization, with similar properties as for perfect synchronization, are the most promising types of synchronization for successful design of our chaos communication scheme. However, we note that sophisticated chaos communication systems may also benefit from the phenomena that are associated with weaker forms of synchronization.

The synchronization properties of the system also determine the qualities of the encryption/decryption process. The reason for this is that the decryption mechanisms of the message of chaos communication systems are based on synchronization phenomena. Therefore, the required synchronization qualities also depend on the method which is used for encryption and decryption of the message. Vice versa, the achievable synchronization quality determines which of the many encryption/decryption methods can be successfully accomplished. For this reason, it is essential to consider the properties of the applied encryption/decryption method for the design of a chaos cryptosystem, as it enters into the overall performance of the system.

Encryption/Decryption Methods

Up to now, several message encryption/decryption techniques have been proposed. The techniques essentially differ in the manner how and where the message is added to the transmitter system. The corresponding decryption methods need to be capable of reconstructing the encrypted message at the receiver. The most famous schemes are *Chaos Masking* (CM), *Chaos Modulation Keying* (CMK), *Chaos Shift Keying* (CSK), and *ON/OFF Shift Keying* (OOSK) [172, 194]. These schemes exhibit inherently different properties with respect to performance and security aspects.

Chaos Masking represents a very simple encryption/decryption method, since the message is only added to the chaotic carrier signal. Therefore, the message is rather concealed than being truly encrypted in this scheme. In the chaos modulation keying scheme, the carrier signal is modulated by the message which only provide a very weak level of encryption [78, 79]. A more sophisticated way to encrypt a message is given by the chaos shift keying method. In CSK, the chaotic dynamics on the strange attractor is directly influenced by a small change of one (or several) of the system parameters of the transmitter. In practice, this is done by switching the parameter between two defined states which causes a shift of the dynamics representing the two states of a bit of binary data. Since the dynamical state is modulated during its nonlinear generation process, the message is nonlinearly mixed into the carrier signal which allows for very efficient encryption [113]. ON/OFF shift keying, on the other hand, represents a principally different scheme, since it is based on switching between synchronization and desynchronization [176]. These two states define the binary states of a bit required for encryption of messages in OOSK.

These encryption/decryption methods reveal different characteristics concerning their masking efficiency, maximum transmission rate and practicability. An ideal encryption/decryption method should meet the following requirements:

- The encryption method should allow for a high level of security. Efficient masking of the message is required so that the message cannot be intercepted by filtering of the transmitted signal in the time or frequency domain.

- The encryption/decryption technique needs to facilitate high transmission rates. Transmission rates in the order of Gbit/s are desired.

- The decryption method needs to be reliable. Good discrimination between the message and the carrier signal is essential for message recovery.

- Structurally simple encryption/decryption methods are preferred for implementation if they also allow for high levels of security.

Not all of the above-introduced encryption/decryption schemes fulfill all these requirements. For instance, chaos masking and chaos modulation keying schemes can be realized without much technical effort. However, the encryption properties of these two schemes are less secure than that of chaos shift keying, which in turn is technically more demanding. Thus, the decryption properties of the different schemes exhibit distinct differences with respect to security, reliability, and practicability. Therefore, it is essential to select the encryption/decryption schemes according to the intended application of the chaos communication scheme, which might necessitate a decision representing a tradeoff between practicability and security.

We have seen that the three principal components of a chaos communication system need to fulfill specific requirements to guarantee functionality of the complete system. The properties of the three components, i.e., the carrier signal, the synchronization quality, and the encryption/decryption method usually cannot be treated and optimized independently. For instance, the bandwidth of the carrier signal usually determines the maximum bandwidth with which a message can be encrypted. In a similar way, the synchronization time in a communication scheme which is based on OOSK also limits the transmission bit rate. Depending on the specific chaos communication scheme, other interdependencies may become of importance that need to be considered for realization of a functional chaos communication systems.

In the following, we explore the possibility of realizing a functional chaos communication system with unidirectionally coupled ECSLs. As a starting point, we investigate possible operation conditions for ECSLs with respect to the generation of suitable chaotic carrier signals. We are looking for emission dynamics which best meets the requirements on the carrier signal properties.

5.2   Generation of Suitable Carrier Signals

The generation of chaotic carriers being suitable for good message encryption with GHz bandwidth is one of the fundamental requirements for building a functional chaos communication system. In this section, we investigate the possibility to generate suitable carrier signals with ECSLs. The results we have obtained in Chapter 3, in which we have analyzed the dynamical properties of chaotically emitting ECSLs, serve as basis for this project. Therefore, we refer to Table 3.2 which provides an overview over key properties of the dynamics of chaotically emitting ECSLs operating in the long and in the short cavity regime. Since the analysis of the dynamical properties revealed that both regimes allow for generation of high-dimensional broadband chaotic intensity dynamics, we consider both, the LCR and the SCR dynamics, as possible candidates. Thus, we analyze the potential of both for the generation of good carrier signals. At first, we address the carrier signal properties of chaotic dynamics generated in the long cavity regime, which will be followed by analysis of the carrier signal properties of chaotic short cavity dynamics.

5.2.1   Carrier Properties in the Long Cavity Regime

Before we evaluate the suitability of chaotic LCR dynamics for chaos communications, we briefly summarize relevant characteristics of the dynamics. In Section 3.4, we have found well-controllable chaotic dynamics with maximum bandwidths in the GHz-range for moderate feedback and for pumping well above the solitary laser threshold. For these conditions ECSLs operate in the fully developed coherence collapse regime (CC). It has also turned out that the bandwidth of the chaotic CC dynamics in the LCR is limited by the relaxation oscillation frequency being typically less than 10 GHz. The rf spectra of the dynamics in this regime showed remains of characteristic system frequencies which could be related to the inverse of the delay time. Additionally, the auto-correlation properties of the dynamics exhibited considerable long-range correlations of up to several nanoseconds as a result of the long delay time. We have demonstrated that the Kaplan-Yorke dimension d_{KY} of chaotic dynamics in the LCR

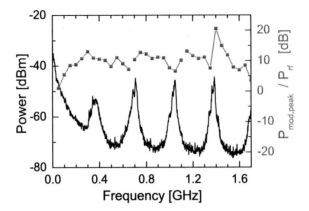

Fig. 5.1: Frequency response characteristics of the chaotically emitting HLP1400 ECSL with
modulation (gray line with symbols) and the corresponding rf spectrum of the
unperturbed chaotically emitting ECSL for $L_{EC} = 40\,$cm, $p = 1.01$ and $r = 0.05$.

is very large, typically $d_{KY} \gg 10$, depending on the length of the external cavity[3],
while the estimation of the Kolmogorov-Sinai entropy h_{KS} of the dynamics gave rise
to moderate values in the LCR.

These dynamical properties represent fairly good carrier signal properties. However, we
have not considered the response characteristics of the dynamics so far. The response
characteristics of the dynamics to small perturbations can give insight into relevant
system properties. Even more, it is possible to decide from the response characteristics
whether the carrier is suitable for realization of the CSK and OOSK methods, since in
these methods, the message is introduced in form of a perturbation of the dynamics.
The signal is not suited for application of these two effective encryption schemes, if
the response characteristics discloses distinct dependencies on the frequency of the
perturbation, or if the dynamics is even insensitive to perturbations.

To characterize the suitability of the potential carrier signal for CSK and OOSK for this
configuration, we measure the response characteristics of the HLP1400 ECSL for LFF
dynamics in the LCR. In the experiment, we adjust for similar conditions as for the
dynamics presented in Figure 3.9. The SL is subject to moderate feedback strength
of $r = 0.05$ with a delay of $\tau_{EC} = 2.7\,$ns. We choose a weak pump parameter of
$p = 1.01$ to better reveal the characteristics dependencies of the modulation response.

---

[3]The dimension linearly grows for increasing the delay time.

The rf spectrum of the dynamics of the ECSLs operating under these conditions is presented by the black solid line in Figure 5.1, which discloses typical features of LFF dynamics, i.e., low frequency components and peaks which are related to the cavity round trip frequency. Then, we apply a perturbation in form of a small harmonic rf modulation of $\Delta p_{mod} \approx 0.03$ to the pump current of the ECSL. By sweeping the frequency of the rf modulation, we were able to determine the frequency response characteristics of the carrier signal which gives insight into the accessible bandwidth of the signal to be encrypted.[4] We measure the response as the difference between the measured peak power of the modulation and the power of the dynamics without modulation at the respective frequency. In the following, we refer to this measure as modulation index. The gray line with the squares, which is presented in Figure 5.1, shows the obtained frequency response characteristics. We notice that the dynamics is only weakly susceptible to modulation at low frequencies. For higher frequencies, the modulation index increases to approximately 10 dB. We note that the modulation index can be increased to 15 dB, if we increase the modulation depth to $\Delta p_{mod} \approx 0.05$. This is the maximal achievable modulation index, since for $\Delta p_{mod} > 0.05$, the dynamics starts to reveal disturbances for frequencies deviating from the modulation frequency. Interestingly, we also identify variations of the modulation index in the vicinity of the cavity round trip frequency of $\nu_{EC} = 375$ MHz and its harmonics. The modulation index reduces for slightly slower frequencies than $\nu = N \times \nu_{EC}$ (with N being an integer), while it is higher for slightly faster frequencies than $\nu = N \times \nu_{EC}$, at the right flank of the peaks.

Our results point out that the modulation response in the long cavity regime is not flat, but reveals some characteristic dependencies. The observed weaker modulation for low frequency dynamics can be explained via the intermittent intensity dropouts in the LFF regime. During these dropouts, the SL emits low power. Thus, the SL less insensitive to the external perturbations. The low modulation index for weak pumping is not problematic regarding the carrier signal properties, since it can be overcome by increasing the pump parameter and going to the fully developed coherence collapse regime, the regime of our interest. However, we have also demonstrated that the carrier modulation index is influenced considerably by the delay. This property might disturb the encryption of real messages which are intrinsically broadband. If such disturbance needs to be avoided, the bandwidth of the message might be adapted to the carrier signal, which in turn reduces the bit rate that can be achieved. Therefore, the results indicate that chaotic CC dynamics of ECSLs operating in the LCR represent acceptable carrier signals with bandwidths in the GHz range. Nevertheless, to come to a final

---

[4]We note that effects on the measured modulation characteristics caused by impedance mismatch between the signal generator and the SL are negligible for modulation frequencies of $\nu_{mod} < 1.4$ GHz.

Physical Properties	
Performance	bandwidth: up to ν_{RO} ($\sim 10\,\text{GHz}$)
	well-controllable parameters
	carrier modulation-index:
	up to 15 dB dependent on ν_{mod}
Privacy	bandwidth: DC to ν_{RO}
	system characteristics in rf spectrum
	correlation $> 3\,\tau_{EC}$ ($>> 1\,\text{ns}$)
	strong nonlinearity $\alpha \approx 3$
	$d_{KY} \gg 10$
	moderate entropy (saturated)
	encryption methods: CSK, CM, CMK
	selective on parameter tuning

Table 5.2: Typical physical properties of carrier signals which can be achieved in the LCR.

decision whether LCR dynamics is suitable carrier signals for chaos communications, we have to consider the other properties of the dynamics as well.

Table 5.2 summarizes the best carrier properties that can be achieved for dynamics in the LCR. These have been found for moderate feedback and moderate pumping conditions, for which ECSLs operate in the fully developed coherence collapse. For these conditions, we have found well-controllable high-dimensional chaotic dynamics with an accessible bandwidth in the GHz range. This underlines that CC dynamics in the LFF regime indeed exhibits suitable carrier signal properties. Nevertheless, the dynamics in the LCR also revealed inherent limitations regarding security and applicability. One of the weak points of LCR dynamics relies in the long auto-correlation times of the dynamics. This property might reveal a potential lack for the security of the system, since it could allow for application of attacks that are based on phase-space reconstruction of the attractor of the dynamics [178,179]. In this context, the residuals of system-inherent cavity round trip frequency in the rf spectra of the dynamics can also be problematic, as they simplify an estimation of the delay time being one of the essential system parameters. Furthermore, it has turned out that the amplitude modulation response depends on the frequency of modulation. This frequency dependence can result in distortions of a message, if the message is encrypted into the carrier signal via modulation of the pump current. As we will show, these and other restrictions can be overcome for ECSLs operating in the short cavity regime. In Chapter 3, we have demonstrated that for delay times being comparable to the period of the relaxation oscillation frequency or shorter, the dynamical properties of the systems qualitatively change if compared to those of the LCR. Therefore, we also study in how far the dynamical properties of ECSLs operating in the SCR can serve as carrier signals for chaos communications.

5.2.2 Carrier Properties in the Short Cavity Regime

In analogy to the analysis of the carrier properties in the LCR, in this subsection, we study the properties of chaotic dynamics generated in the SCR regime and discuss whether the dynamics can also be used as carrier signals for chaos communications. Again, we refer to the results we have obtained for the detailed characterization of chaotic SCR dynamics which we have presented in Section 3.4, while the main results are summarized in Table 3.2.

In contrast to the robustness of the dynamics in the LCR, in the SCR special care has to be taken for the stability of the system, because of the sensitivity of the dynamics on the feedback phase $\Delta\Phi$. However, we have demonstrated that this technical problem can be solved by design of robust ECSL setups which allow for good stabilization and control of this relevant system parameter. Then, $\Delta\Phi$ represents a good control parameter for the dynamics. For carefully adjusted $\Delta\Phi$, it is possible to achieve excellent chaotic dynamics in the SCR, when we also adjust for moderate feedback strength and for pump parameters well above the solitary laser threshold. For these conditions, we have demonstrated broadband chaotic intensity dynamics with a continuous and almost flat rf spectrum up to the first external cavity resonance of 3.85 GHz, as depicted in Figure 3.22 b). The auto-correlation properties of the chaotic SCR dynamics give rise to fall-off of the correlations within only 2 to 3 round trips of the light in the external cavity, corresponding to a short correlation range of about 400 ps. This fast decaying correlation prevents successful application of embedding-techniques for phase-space reconstruction of the chaotic attractor [178, 179]. We note that the bandwidth of the dynamics can easily be enhanced to frequencies well beyond 10 GHz by reducing the length of the external cavity [86]. Nevertheless, our numerical findings revealed that a minimum external cavity length of several centimeters is required to guarantee the desired high-dimensional chaotic dynamics with a Kaplan-Yorke dimension of $d_{KS} \approx 10$. In the same range of the external cavity length L_{EC}, chaotic SCR dynamics also exhibits maximum Kolmogorov-Sinai entropy h_{KS}. We have determined the maximum of h_{KS} for $L_{EC} \approx 3$ cm for the considered laser parameters. Hence, the available bandwidth for optimal operation conditions of the investigated ECSL amounts to approximately $5\,\mathrm{GHz}$. In this context, we point out that the optimal bandwidth of the dynamics also depends on the laser parameters, particularly on the α-parameter, which allows for further extension of the bandwidth by selection of SLs with optimized parameters. Nevertheless, the realized dynamics already reveals excellent carrier signal properties, comprising broad bandwidth, high dimensionality, large entropy, and short-range correlation. However, we also need to verify whether the dynamics is suitable for realization of sophisticated message encryption techniques, such as CSK and OOSK.

To investigate the suitability of chaotic SCR dynamics for implementation of CSK, again we measure the modulation response characteristics of the dynamics. Therefore, we choose operation conditions which are close to those for best carrier signal properties. In the experiment, the length of the external cavity amounts to $L_{EC} = 3.9\,\mathrm{cm}$, which corresponds to $\nu_{EC} = 3.84\,\mathrm{GHz}$. We apply moderate feedback to the SL with $r = 0.05$. Only the pump parameter considerably deviates from the optimum value of about $p = 1.50$. In similarity to the experiment in the LCR, we reduce the pump parameter to better reveal characteristic frequencies in the dynamics and in the modulation response. Therefore, we adjust for $p = 1.20$ for which the relaxation oscillation frequency of the solitary laser has been determined to be $\nu_{RO} \approx 2.3\,\mathrm{GHz}$. We are able to obtain broadband chaotic dynamics for carefully adjusted feedback phase $\Delta\Phi$, while corresponding numerical results verify high dimensional dynamics of $d_{KY} \approx 10$.

The high spectral bandwidth of the obtained chaotic intensity dynamics can be deduced from Figure 5.2, in which the detection noise level corresponds to $-82\,\mathrm{dBm}$. Please note that the depicted rf spectrum already indicates results for the modulation experiments, while the rf spectrum of the ECSLs without perturbations corresponds to the presented spectrum without the three modulation peaks located at $\nu_1 = 0.5\,\mathrm{GHz}$, $\nu_2 = 1.7\,\mathrm{GHz}$, and $\nu_3 = 2.5\,\mathrm{GHz}$. The rf spectrum reveals the high bandwidth of the dynamics of about $4\,\mathrm{GHz}$. For this low level of pumping, the dynamics give rise to two characteristic frequencies appearing as broad humps in the spectrum close to the external cavity round trip frequency $\nu_{EC} = 3.85\,\mathrm{GHz}$ and close to the relaxation oscillation frequency of the solitary SL $\nu_{RO} = 2.3\,\mathrm{GHz}$. Besides this, the spectrum represents a continuous and flat distribution of the power of the frequency components being involved in the dynamics.

For these conditions we have applied CSK by adding a small harmonic current modulation of $\Delta p_{mod} = 0.02$ to the DC bias current of the SL so that the rf spectrum does not exhibit perturbations for frequencies differing from ν_{mod}. The three sharp peaks in the rf spectrum in Figure 5.2 originate from three different modulation response experiments for $\nu_1 = 0.5\,\mathrm{GHz}$, $\nu_2 = 1.7\,\mathrm{GHz}$, and $\nu_3 = 2.5\,\mathrm{GHz}$, respectively. The results disclose similar modulation indices for the three different modulation frequencies. In a continuative experiment, we sweep the modulation frequency within a range of $100\,\mathrm{MHz}$ and $5\,\mathrm{GHz}$ and find that the modulation index for frequencies between

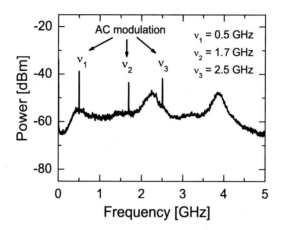

Fig. 5.2: Frequency response of the HLP1400 semiconductor laser operated in the short cavity regime under following conditions: $\nu_{EC} = 3.84\,\text{GHz}$, $p = 1.20$ and $r = 0.05$. The arrows in the figure mark three frequencies of modulation.

$100\,\text{MHz} < \nu_{mod} < 4\,\text{GHz}$ can reach up to 20 dB. The measured modulation characteristic exhibits variations of only $\approx 5\,\text{dB}$ in this spectral range. However, we can identify systematic changes of the modulation index. Firstly, it slightly increases for frequencies close to $\nu_{RO} = 2.3\,\text{GHz}$. Secondly, the modulation index continuously and slowly decreases for increasing the modulation frequency up to 4 GHz, while it drastically reduces for frequencies beyond 4 GHz. This observed systematically decreasing modulation index for frequencies beyond 2 GHz can be attributed to the electrical properties of the SL mount, which has not been designed for high-frequency modulation. These disturbing effect can be easily avoided by using SL which are designed for rf modulation. Therefore, we highlight that the response characteristics of chaotic SCR dynamics exhibit excellent properties in a wide range of frequencies only limited by the external cavity round trip frequency $\nu_{EC} = 4\,\text{GHz}$.

In comparison to the requirements defined in Table 5.1, we summarize the characteristics of chaotic SCR dynamics in Table 5.3. It becomes clear from the table that the privacy-properties are complemented by good performance-properties. These are the high modulation index, indicating a good signal-to-noise ratio of the carrier, the good sensitivity on parameter tuning and the robustness against environmental influence. Furthermore, the SCR carrier generator can be implemented in a compact and cost-effective way in a chaos communication system. In conclusion, the requirements on the

Physical	
Performance	bandwidth: up to ν_{EC} ($\gg 10\,\text{GHz}$ possible)
	well-controllable parameters
	carrier modulation-index:
	20 dB for ν_{mod} up to ν_{EC}
Privacy	bandwidth: DC to ν_{EC} ($\nu_{EC} > \nu_{RO}$)
	broad, almost flat rf spectrum
	short correlation $< 3\,\tau_{EC}$ ($\ll 1$ ns)
	strong nonlinearity $\alpha \approx 3$
	$d_{KY} \approx 10$
	high entropy (maximum in SCR)
	encryption methods: CSK, CM, CMK
	selective on parameter tuning

Table 5.3: Typical physical properties of carrier signals which can be realized in the SCR.

carrier signal generator defined in Table 5.1 can be entirely fulfilled in the SCR.

In the next step towards realization of a chaos communication system, we need to study whether a receiver system can synchronize to the carrier signal of the transmitter. Therefore, we present two different receiver configurations, the closed loop and the open loop configuration, and investigate their potential for realization of chaos synchronization. We still consider both LCR and SCR carrier signals, keeping in mind that chaotic SCR dynamics exhibits better carrier signal properties than chaotic LCR dynamics. Nevertheless, the robustness of the dynamics in the LCR regime and its limited bandwidth sightly simplify the high demands on the challenging synchronization experiments. In that sense, first chaos synchronization experiments in the LCR can already give insight into relevant synchronization properties. Therefore, they can serve as basis for subsequent more sophisticated synchronization experiments in the SCR.

5.3 Synchronization Properties: The Closed Loop Configuration

In this section, we study the possibility of synchronizing a receiver system to the chaotic dynamics (carrier signal) of the transmitter system. In the chosen configuration, the transmitter and the receiver are similar systems. Each system consists of a chaotically emitting ECSL, while the SLs and the system parameters of the two systems need to be well-matched. The ECSLs are unidirectional coupled via injection of the dominant TE component of the electromagnetic field of the carrier signal into the receiver system. This configuration is generally referred to as *Closed Loop Configuration*, which is contrasted to the simpler open loop configuration in which the receiver only consists of a solitary SL.

The possibility of chaos synchronization in the closed loop configuration has been first addressed by modeling [78, 134]. Just five years ago, Fujino et al. reported first successful experimental synchronization of LFF dynamics in the closed loop configuration [183]. Since then, the synchronization properties of this configuration have attracted much interest in particular the LFF regime has been intensively studied. This has been done numerically [195–198] and experimentally [49, 144, 189, 199, 200]. However, up to now, detailed studies on the synchronization properties in the fully developed chaotic regime and its dependence of the systems parameters have been lacking. This is due to the fact that the experiments in this dynamical regime are increasingly challenging with growing dynamical bandwidth. We address this problem and present comprehensive experimental and numerical investigations of the synchronization properties of two unidirectionally coupled ECSLs which both operate in the fully developed chaotic regime. We study the influence of the system parameters on the synchronization behavior of the SLs and reveal fascinating synchronization phenomena. Additionally, we characterize the synchronization properties with respect to applicability for chaos communications. Our results demonstrate that this optical system is well-suited for studying general synchronization phenomena of high-dimensional nonlinear oscillators, since the system parameters are well-controllable and the dynamical properties of chaotically emitting ECSLs are well-understood. Furthermore, we verify the high potential of this configuration for the realization of functional chaos communication systems.

5.3.1 Experimental Setup

A scheme of the experimental setup of the closed loop configuration is depicted in Figure 5.3. The transmitter and the receiver systems consist of two device-identical Hitachi

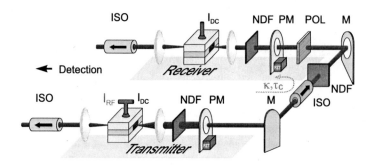

Fig. 5.3: Schematic representation of the closed loop configuration which consists of two
unidirectionally coupled chaotic external cavity semiconductor lasers.

HLP1400 Fabry-Perot SLs with well-agreeing parameters. The frequency characteristics of the optical spectra of the SLs agree to better than 0.1 nm, their slope efficiencies
coincide to better than 3 % and the threshold currents $I_{th,sol}$ deviate by only 7 %. The
SLs are pumped by ultra low-noise DC current sources and are temperature stabilized
to better than 0.01 K. Each SL is subject to delayed optical feedback from a distant
partially reflective mirror (PM). Therefore, the transmitter and the receiver system
represent similar ECSLs. The cavity round trip time τ_{EC} of both system has been
carefully matched. In the following, we refer to the delay time of the transmitter and
the receiver as τ_t and τ_r, which amount to $\tau_t = \tau_r = \tau_{EC} = 2.9$ ns. Here, τ_{EC} denotes
the delay of both systems in the case of identical delays. Additionally, the feedback
ratios $r_{t,r}$ can be adjusted in each of the system by using neutral density filters (NDF).
The feedback phases $\Delta\Phi_{t,r}$ can be independently controlled by changing $L_{EC,t}$ and
$L_{EC,r}$ on sub-wavelength scale using piezo transducers (PZT). The receiver is coupled
to the transmitter by injection of a well-defined fraction of the electromagnetic field of
the transmitter into the receiver. The optical isolator (ISO), and the polarizer (POL)
guarantee unidirectional coupling via the dominant TE component of the electromagnetic field, while the coupling rate κ_c can be adjusted using a neutral density filter
(NDF). The coupling strength can also be expressed by the ratio between the power
injected into the receiver and the power emitted by the transmitter $r_c = P_{inj,r}/P_{out,t}$.
In most of the experiments, the coupling time τ_c amounts to 4.6 ns, though we note
that τ_c is not relevant as the coupling is unidirectional. We detect the intensity dynamics of the transmitter and the receiver simultaneously with two photodetectors
$(PD_{1,2})$ with a bandwidth of either 6 GHz or 12 GHz, respectively. In this way, we can
capture the fast dynamics on sub-ns time scale, which we acquire with a fast digital

oscilloscope with 4 GHz analog bandwidth and a sampling rate of 10 GSamples/s on each channel. The spectral properties of the intensity dynamics can be studied by help of an electrical spectrum analyzer (ESA) with 21 GHz bandwidth. Furthermore, we monitor the optical spectra using an optical grating spectrum analyzer (OSA) with 0.05 nm resolution.

5.3.2   Synchronization in the Long Cavity Regime

In the following, we verify that synchronization of chaotic LCR dynamics of the transmitter and the receiver is possible in the closed loop scheme. In subsequent experiments, we study the synchronization properties in dependence of the relevant system parameters. We demonstrate that the synchronization properties in the closed loop configuration strongly depend on the operation parameters and in particular on the amplitude and phase of the coupling. Our experiments are substantiated by numerical simulations which verify that the synchronization scenario is closely linked to the vectorial character of the coupling.

The Low Frequency Fluctuations Regime

Initially, we choose the experimental conditions such that the transmitter and receiver operate in the well-studied LFF regime for a pumping parameter of $p_{t,r} = 1.01$, before we study the synchronization properties in the fully developed CC regime for $p_{t,r} = 1.16$. In the LFF regime, we carefully match the transmitter and the receiver dynamics to achieve synchronization of both systems. We adjust the strength of the delayed optical feedback in the transmitter and in the receiver using the neutral density filters (NDF) such that we measure a threshold reduction of $\Delta r_t = -0.07$ for the transmitter SL and $\Delta r_r = -0.04$ for the receiver SL without coupling. These values correspond to feedback ratios of $r_t = 0.07$ and $r_r = 0.05$. Additionally, it is necessary to match the optical frequencies of the transmitter and the receiver SL to better than 1 GHz, which can be realized by appropriate control of the temperature of the SLs. In agreement with [183], we find that best synchronization can be achieved when the sum of coupling and feedback intensity in the receiver is slightly larger than the feedback intensity in the transmitter. Indeed, after optimization of the feedback and the coupling conditions, and precisely matching $\Delta\Phi_{t,r}$, we are able to achieve chaos synchronization in the LFF regime. We assign the relative optical feedback phase being $\Delta\Phi_{rel} = 0$ to these conditions[5], a choice to be supported later by our numerical modeling results.

---

[5] We note that $\Delta\Phi_{t,r}$ is well defined for the transmitter and the receiver system, as we have adjusted for $L_{EC,t,r} = N \times n_{eff}L_{SL}$, where N represents an integer, while n_{eff} stands for the effective

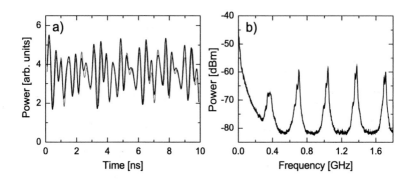

Fig. 5.4: Panel a), time series of the intensity dynamics of the transmitter (gray) and the receiver (black) arranged in the closed loop configuration and operating in the LFF regime for $p_{t,r} = 1.01$. The time series of the receiver has been shifted by the coupling time $\tau_c = 4.6\,\text{ns}$. The corresponding rf spectra are depicted in panel b).

Figure 5.4 depicts the dynamics of the transmitter and the receiver for optimized synchronization, which we find for the following combination of feedback and coupling ratios: $r_t = 0.07$, $r_r = 0.05$ and $r_c = 0.03$. In Figure 5.4 a), we illustrate a 10 ns long segment of the intensity dynamics. The dynamics of the transmitter are represented in gray, while that of the receiver is represented in black. In the experiment, we measure that the receiver dynamics lags the transmitter dynamics by $\Delta t = 4.6\,\text{ns}$, which exactly corresponds to the coupling time of $\tau_c = 4.6\,\text{ns}$. Accordingly, as indicated by this fixed time lag, and the non-symmetrical feedback and coupling conditions, we do not observe identical synchronization here, but generalized synchronization [201]. We note that we do not find the perfect synchronization solution our the experiments, which predicts a delay of the receiver of $\Delta t = \tau_c - \tau_{EC}$ [200]. Nevertheless, the realized generalized synchronization exhibited excellent correlation of the amplitude and phase of the intensity dynamics of both systems. This becomes clear in Figure 5.4 a), in which we have shifted the time series of the receiver dynamics by τ_c to compensate for its delay. The comparison of the time series discloses almost perfectly agreeing dynamics down to the relevant sub-ns time scale of the dynamics, which manifests itself in long-time-averaged cross-correlation coefficients of up to $\Sigma_{corr} = 0.9$. We point out that the

---

refractive index of the gain material of the SLs with cavity length L_{SL}. For these conditions, $\Delta\Phi_{t,r}$ is the same for all of the longitudinal modes of the transmitter and the receiver SL. Furthermore, we have numerically estimated that the maximum phase deviation that occurs between the lasing modes for a phase shift of $\Delta\Phi = 10\pi$ only amounts to 0.04π and, thus can be neglected.

synchronization is robust against reasonable variations of the pump parameter and the feedback strength.

To provide complementary insight into the synchronization quality, in Figure 5.4 b) we also depict the corresponding rf spectra of the transmitter and the receiver dynamics. The almost perfectly coinciding rf spectra reveal synchronization of both, the fast sub-ns intensity pulsations and the characteristic low frequency intensity dropouts. These findings underline the excellent synchronization quality in the LFF regime, which is confirmed by the well-agreeing optical spectra of the dynamics. Hence, we have verified that it is principally possible to synchronize high-dimensional chaotic intensity dynamics of ECSLs. In a next step, we are interested in whether synchronization can also be achieved in the fully developed coherence collapse regime, in which the bandwidth of the dynamics is considerably higher than in the LFF regime.

The Fully Developed Coherence Collapse Regime

The transition from LFF dynamics to fully developed coherence collapse dynamics is accompanied by considerable increasing bandwidth of the dynamics. Reason for this can be found in the increasing relaxation oscillation frequency and the enhancing irregularity of the dynamics, manifesting itself in growing Kolmogorov-Sinai entropy of the dynamics [140]. In the next experiment, we study the influence of enhancing spectral bandwidth of the dynamics on the synchronization properties in the closed loop configuration. Therefore, we simultaneously increase the pump parameters of both, the transmitter and the receiver SL, to $p_{t,r} = 1.16$. The remaining parameters are kept constant. We only slightly readjust the relative feedback phase to $\Delta\Phi_{rel} = 0$ so that we achieve best possible synchronization.

Figure 5.5 depicts the corresponding intensity dynamics of both coupled ECSLs for fully developed CC dynamics. In panel a) we present the time series for optimized synchronization between the transmitter and the receiver. The dynamics of the transmitter is always presented in gray, while that of the receiver is depicted in black. The time series show good agreement between the transmitter and the receiver dynamics. To allow for simple comparison of the dynamics, we again compensate for the delay of the receiver dynamics which lags by $\tau_c = 4.4\,\text{ns}$. Then, we can easily determine the cross-correlation coefficient of the dynamics which amounts to $\Sigma_{corr} = 0.85$. Such high correlation demonstrates that it is indeed possible to realize excellent synchronization in the CC regime. Further investigating the synchronization, we monitor the rf spectra and the optical spectra of the transmitter and the receiver. The corresponding rf spectra are illustrated in Figure 5.5 b). The rf spectra disclose excellent agreement for frequencies of up to 1 GHz, and show good qualitative agreement for higher fre-

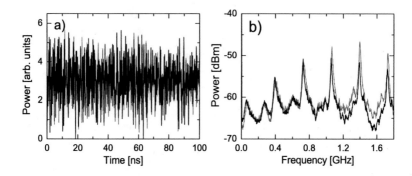

Fig. 5.5: Panel a), time series of the intensity dynamics of the transmitter (gray) and the receiver (black) arranged in the closed loop configuration and operating in the coherence collapse regime for $p_{t,r} = 1.16$. The time series of the receiver system has been shifted by the coupling time $\tau_c = 4.6$ ns. The corresponding rf spectrum is depicted in panel b).

quency components. Furthermore, we find that the optical (multimode) spectra for CC conditions almost perfectly coincide, confirming the good synchronization. In the experiment, we have slightly varied the pumping parameter, the coupling, and the feedback strength and find that the synchronization is robust against variations of the order of 5 %. Nevertheless, the results also disclose that synchronization in the CC regime is more susceptible to variations of the system parameters than synchronization in the LFF regime.

These findings clearly demonstrate that it is possible to realize generalized synchronization with good synchronization quality in the CC regime. Nevertheless, the synchronization quality is slightly reduced and more sensitive to variations of system parameters, if compared to the synchronization quality we have achieved in the LFF regime. Therefore, these results also indicate the increasing demands on the experiments for increasing the bandwidth of the dynamics. To achieve good synchronization of high-dimensional broadband dynamics, it is essential to select very well matching SLs and to precisely match the operation conditions of the transmitter and the receiver system. Additionally, the experimental setup needs to meet high requirements on its mechanical robustness, which becomes clear from the following experiment, in which we vary the relative cavity phase.

5.3.3   Scenario in Dependence on the Relative Cavity Phase

In this section, we study the sensitivity of the synchronization in the CC regime in the closed loop configuration on the relative feedback phase $\Delta\Phi_{rel} = \Delta\Phi_t - \Delta\Phi_r$. Our motivation to study the influence of this parameter arises from the peculiar type of coupling that is realized in the closed loop configuration. In this optical system, the transmitter and the receiver are coupled via their electromagnetic fields. Therefore, the coupling parameter is a two-component vector consisting of the field amplitude and the optical phase. The influence of the coupling field amplitude, i.e., the coupling strength, on the synchronization behavior of the system has been studied numerically in [195], and experimentally for the open loop scheme [188]. However, the influence of the relative feedback phase has been neglected in these previous investigations. Only recently, we have shown that the optical feedback phase cannot be neglected when studying synchronization phenomena in the closed loop configuration [144]. Even more, the relative feedback phase represents one of the key parameters because of the vectorial nature of the coupling. In this previous work, we have demonstrated experimentally and numerically that for LFF dynamics a controlled variation of the relative optical feedback phase $\Delta\Phi_{rel}$ leads to a dynamical synchronization scenario. In this scenario, the receiver dynamics varies between chaos synchronization, and weakly correlated states. For adjusted phase, it was possible to achieve excellent synchronization of the intensity dynamics of the transmitter and the receiver in conjunction with coherence of the optical fields. Variation of $\Delta\Phi_{rel}$ led to conspicuous changes of the receiver dynamics associated with a drastically reduced correlation and loss of the coherence of the optical fields. Finally, synchronization could be regained for a phase shift of $\Delta\Phi_{rel} = 2\pi$, underlining the cyclic nature of the control parameter. This demonstration of the key relevance of $\Delta\Phi_{rel}$ for the synchronization behavior of the system was particularly remarkable since the dynamics of solitary ECSLs operating in the LCR does not give rise to changes of the temporal and spectral properties of the dynamics for variation of $\Delta\Phi_{rel}$. However, in closed loop configuration the receiver can act as a very sensitive detector for a slight change of the feedback phase of the transmitter. This is due to the fact that even a small variation of $\Delta\Phi_t$ changes the synchronization properties of the system. Since the previous investigations were restricted to synchronization phenomenon in the LFF regime, we need to study the influence of $\Delta\Phi_{rel}$ on the synchronization properties in the CC regime, in which the dynamics exhibits better properties with respect to chaos communications. We choose the same experimental conditions as for the previous experiment. We select the state of best synchronization as starting point from which we detune the feedback phase. For this condition the relative feedback phase amounts to $\Delta\Phi_{rel} = 0$. Figure 5.6 a) depicts a 10 ns long segment of the intensity dynamics for synchronization. For this state, the time series reveal good

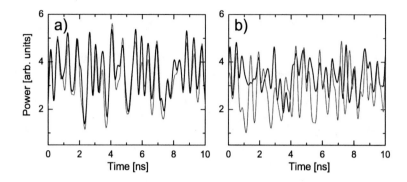

Fig. 5.6: Synchronization behavior of the transmitter dynamics (gray line) and the receiver
dynamics (black line) in dependence of the relative optical feedback phase $\Delta\Phi_{rel}$;
Panel a) CC operation with $p_{t,r} = 1.16$ and adjusted phase $\Delta\Phi_{rel} = 0$. Panel b)
CC operation for the same pump parameter, but with $\Delta\Phi_{rel} = 0.7\pi$.

synchronization of the amplitude and phase of the irregular oscillations. However, Figure 5.6 b) demonstrates that the receiver dynamics drastically changes when we set the
relative feedback phase to $\Delta\Phi_{rel} = 0.7\pi$ by changing $\Delta\Phi_t$. Then, synchronization is
lost in the CC regime. Now, the time series, the rf spectrum, and the optical spectrum
of the receiver clearly differ to that of the transmitter. The dynamical properties of the
transmitter, of course, remain unaffected due to the undirectional coupling. This result
demonstrates that synchronization of CC dynamics in the closed loop configuration is
also very sensitive to the relative feedback phase.

In the following, we investigate the transition from synchronization to uncorrelated
states in detail. We focus on the synchronization behavior of the system for continuous
variation of the relative feedback phase $\Delta\Phi_{rel}$. We note that this parameter can either
be changed by variation of $\Delta\Phi_r$ or by $\Delta\Phi_t$. Without loss of generality, we restrict the
experiment to changing $\Delta\Phi_{rel}$ via $\Delta\Phi_t$, as this would also be done for realization of
OOSK in a possible chaos communication system.

In the experiments, we find that as soon as $\Delta\Phi_{rel}$ deviates significantly from $\Delta\Phi_{rel} = 0$
a sudden switching behavior sets in, in which the receiver dynamics mediates between
synchronized states with highly correlated dynamics of the two SLs, and desynchronized
states with low correlated dynamics. The mean frequency of the switching increases
for continuously incrementing $\Delta\Phi_{rel}$. As a consequence thereof, the maximum cross-
correlation between the receiver and the transmitter dynamics steadily decreases until

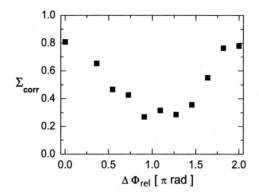

Fig. 5.7: Cross correlation coefficients of the intensity time series of the transmitter and the receiver versus the relative feedback phase $\Delta\Phi_{rel}$ for CC dynamics.

synchronization is lost for $\Delta\Phi_{rel} = \pi$. This means that a gradual detuning of $\Delta\Phi_{rel}$ away from the synchronization condition being $\Delta\Phi_{rel} = 0$ degrades the temporal synchronization behavior of the system. When we continue with increasing $\Delta\Phi_{rel}$, we find that synchronization is slowly regained, until it is fully recovered for $\Delta\Phi_{rel} = 2\pi$. In analogy to the results that have been reported for LFF dynamics, the relative optical feedback phase $\Delta\Phi_{rel}$ turns out to be a 2π-cyclic parameter. To quantify the described behavior, we calculate the maximum time-averaged correlation coefficient Σ_{corr} between the transmitter and the receiver dynamics in dependence of $\Delta\Phi_{rel}$. We use the following definition for estimating the maximum cross-correlation coefficient Σ_{corr} for the experimental data:

$$\Sigma_{corr} = \max_{\Delta t}\{\frac{\langle\delta P_t(t)\delta P_r(t-\Delta t)\rangle}{\sqrt{\langle\delta P_t^2(t)\rangle\langle\delta P_r^2(t-\Delta t)\rangle}}\} \tag{5.1}$$

In this notation, $P_{t,r}(t)$ represents the intensity of the transmitter and the receiver dynamics, and $\delta P_{t,r}(t) = P_{t,r}(t) - \langle P_{t,r}\rangle$. The angled brackets stand for time averaging. According to this definition, the intensity dynamics is best correlated when $\Sigma_{corr} = 1$, while it is completely uncorrelated when $\Sigma_{corr} = 0$. The parameter Δt is introduced to account for eventual time lags between the receiver and the transmitter dynamics, as it occurs for the present generalized synchronization. The cross-correlation analysis shows that the maximum $\Sigma_{corr}(\Delta t)$ is observed for $\Delta t = \tau_c$, as long as $\Sigma_{corr} \geq 0.5$. Hence, $\Sigma_{corr} = 0.5$ can be treated as lower limit for appearance of generalized synchronization.

Figure 5.7 provides an overview over the synchronization scenario in terms of Σ_{corr}

in dependence of $\Delta\Phi_{rel}$. The figure indicates high correlation coefficients for adjusted phase $\Delta\Phi_{rel} \approx 0$, for which we find good synchronization of the intensity dynamics of the transmitter and the receiver. We observe that the correlation slowly decreases for increasing $\Delta\Phi_{rel}$, until minimal correlation is reached at around $1\pi < \Delta\Phi_{rel} < 1.4\pi$. For further increasing $\Delta\Phi_{rel}$, the correlation quickly improves until chaos synchronization is regained for $\Delta\Phi_{rel} \approx 2\pi$. It is important to note, that $\Sigma_{corr}(\Delta\Phi_{rel}$ is asymmetric with respect to $\Delta\Phi_{rel}$, i.e., $\Sigma_{corr}(\Delta\Phi_{rel}) \neq \Sigma_{corr}(2\pi - \Delta\Phi_{rel})$. This asymmetry results from the unidirectional coupling which breaks the symmetry in the system. In the following, we present the corresponding Lang-Kobayashi rate-equation model from which we can gain further insight into the synchronization properties of chaotically emitting ECSLs in the closed loop configuration. From these equations, it can be seen that the unidirectional coupling leads to invariance of the equations against substitution of $\Delta\Phi_{rel}$ by $-\Delta\Phi_{rel}$.

5.3.4   Modeling Results

In Chapter 3, we have demonstrated that the Lang-Kobayashi SL rate-equations are often well-suited for studying the dynamics of chaotically emitting ECSLs. In this section, we benefit from this knowledge and extend the LK equations Eqs. 3.3-3.5 such that we can model the dynamics of the transmitter (t) and the receiver (r) system in the closed loop configuration. The modeling has been performed in collaboration with J. Mulet, P. Colet and C.R. Mirasso from the IMEDEA and the UIB, Palma de Mallorca, Spain. In the model, the dynamics of the transmitter and the receiver system are described by rate-equations for the complex slowly varying amplitude of the electric fields $E_{t,r}(t)$ and the carrier numbers $N_{t,r}(t)$. The adapted rate-equations reflecting the situation in the closed loop configuration read:

$$\dot{E}_t(t) = \frac{1}{2}(1 + i\alpha)\left[G_t(t) - \Gamma\right] E_t(t) + \gamma_t e^{-i(\Phi_t(t) + \Delta\Phi_t)} E_t(t - \tau_t) + \sqrt{2\beta\gamma_N N_t(t)}\,\zeta_t(t)\,, \tag{5.2}$$

$$\dot{E}_r(t) = \frac{1}{2}(1 + i\alpha)\left[G_r(t) - \Gamma\right] E_r(t) + \gamma_r e^{-i(\Phi_r(t) + \Delta\Phi_r)} E_r(t - \tau_r) + \sqrt{2\beta\gamma_N N_r(t)}\,\zeta_r(t) + \kappa_c e^{-i\omega_0\tau_c} E_t(t - \tau_c)\,, \tag{5.3}$$

$$\dot{N}_{t,r}(t) = \frac{p\,I_{t,r}^{th,sol}}{e} - \frac{N_{t,r}(t)}{T_N} - G_{t,r}(t)|E_{t,r}(t)|^2\,, \tag{5.4}$$

$$G_{t,r}(t) = \frac{\xi(N_{t,r}(t) - N_0)}{1 + \varepsilon|E_{t,r}(t)|^2}\,. \tag{5.5}$$

Symbol	Meaning	Value
α	linewidth enhancement factor	5
ξ	differential gain	$6 \times 10^{-6}\,\text{ns}^{-1}$
Γ	cavity decay rate	$200\,\text{ns}^{-1}$
T_N	carrier life time	$0.5\,\text{ns}$
N_0	transparent carrier number	1.5×10^8
ϵ	gain suppression coefficient	5×10^{-7}
β	spontaneous emission factor	10^{-7}
$\tau_{t,r}$	external cavity round trip time	$3\,\text{ns}$
$I_{th,sol}$	solitary threshold current	≈ 60 mA

Table 5.4: Device and operation parameters for modeling.

For simplicity, we assume that both SLs have identical device parameters, so that the emission frequency $\omega_0 = 360\,\text{THz}$ agrees for the same operation conditions. The first terms in the right-hand side of the field Eqs. 5.2 and 5.3 describe the gain-loss balance of the solitary SLs. The second terms account for the delayed optical feedback, with the external cavity round trip times $\tau_{t,r}$ and the feedback rates $\gamma_{t,r}$. The accumulated phase difference of the electric field in one round trip within the external cavity is denoted as $\Delta\Phi_{t,r} = \omega_0\tau_{t,r}$, mod 2π. The last term in Eq. 5.3 describes the unidirectional coupling via injection of the electromagnetic field of the transmitter into the receiver. For unidirectional coupling, it is possible to rescale the local time and the phase in the receiver system such that the coupling time τ_c can be compensated for. Then, we can take $\tau_c = 0$ and $\omega_0\tau_c$ mod $2\pi = 0$ without loss of generality. In the model, we would like to consider conditions which are close to the experiment. Therefore, we also include complex Gaussian noise terms $\zeta_{t,r}(t)$, which we add to the field equations to model spontaneous emission noise. These numbers have zero mean $\langle\zeta_{t,r}(t)\rangle = 0$ and the correlations $\langle\zeta_a(t)\zeta_b^*(t')\rangle = 2\delta_{a,b}\delta(t-t')$, with $(a,b) = t,r$. By adding noise, we are able to numerically verify stability of the synchronization solution. The gain function in the carrier Eq. 5.4 is assumed to depend linearly on the carrier number. We also account for nonlinear gain-suppression effects in Eq. 5.5. Then, in similarity to the procedure we have applied in Chapter 3, we numerically integrate the equations of the model using parameters that correspond to the experimental conditions, which are summarized in Table 5.4.

According to the experimental conditions, we consider matched external cavity lengths for the transmitter and the receiver system. Therefore, we set $\tau_{EC} = \tau_t = \tau_r$, while the relative feedback phase can be varied within $\Delta\Phi_{rel} = \Delta\Phi_t - \Delta\Phi_r \in [0, 2\pi]$. For the sake of simplicity and numerical purposes, it is convenient to introduce the following rescaling of the dynamical variables:

$$A_{t,r}(t) = \sqrt{\frac{\xi}{\gamma_N}} E_{t,r}(t), \qquad D_{t,r}(t) = \frac{N_{t,r}(t)}{N_0} - 1. \tag{5.6}$$

As expected, in the numerical simulations, we find that in the absence of coupling, for $\kappa_c = 0$, the intensity dynamics of both lasers are completely uncorrelated. When we continuously increase κ_c, we identify a critical value of coupling beyond which synchronization effects can be observed. This value corresponds to $\kappa_c \approx 15\,\mathrm{ns}^{-1}$ for the considered set of parameters. In agreement with the experiments, we find that the correlation of the dynamics of the transmitter and the receiver system is not only influenced by the coupling strength, but even more it exhibits strong dependence of the relative feedback phase. This dependence is illustrated in Figure 5.8, in which panel a) depicts segments of the intensity dynamics of the transmitter and the receiver for operation in the CC regime. The dynamics of the transmitter is represented in gray, while the dynamics of the receiver is depicted in black. In this case, we consider moderate feedback and coupling of $\gamma_t = \gamma_r = \kappa_c = 20\,\mathrm{ns}^{-1}$. The pump parameters are set to $p_{t,r} = 1.50$. The upper figure in panel a) reveals high correlation of the intensity dynamics of the transmitter and the receiver for adjusted feedback phase $\Delta\Phi_{rel} = 0$. In agreement with the experiments, we find generalized synchronization for which the time delay between the dynamics of the transmitter and the receiver system corresponds to τ_c, which has been set to $\tau_c = 0$ in the modeling. This situation drastically changes in the lower figure of panel a), in which we have shifted the feedback phase of the transmitter such that $\Delta\Phi_{rel} = \pi$. For this phase condition, we can see that the correlation between the transmitter and the receiver dynamics is lost, which is in good agreement with our experimental results presented in Figure 5.6 b), underlining the suitability of the model. Furthermore, it demonstrates the importance of $\Delta\Phi_{rel}$ for the synchronization behavior of the system.

Next, we study the synchronization behavior for variation of $\Delta\Phi_{rel}$ in the interval $[0, 2\pi]$. We also quantify the quality of the synchronization in terms of the cross-correlation coefficient as defined in Eq. 5.1. This allows for direct comparison of experiment and numerical modeling. The numerical results are presented in Figure 5.8, panel b) which provides an overview over the synchronization scenario in terms of Σ_{corr} as a function of $\Delta\Phi_{rel}$, for operation in the CC regime. For selection of similar parameters as for the experimental conditions, i.e., moderate coupling and slightly less feedback in the receiver system than in the transmitter system, we find good qualitative agreement with the experiment. This can be seen from the black line in panel b), which presents results for $\kappa_c = \gamma_t = 20\,\mathrm{ns}^{-1}$, $\gamma_r = 10\,\mathrm{ns}^{-1}$, and $p = 1.50$. For these

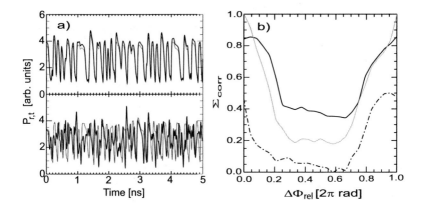

Fig. 5.8: Panel a) presents numerically obtained intensity time series ($P_{t,r} = |A_{t,r}|^2$) for dynamics in the CC regime $p_{t,r} = 1.5$. The dynamics of the transmitter timer is depicted in gray and that of the receiver in black. The parameters correspond to: $\gamma_t = \gamma_r = \kappa_c = 20\,\mathrm{ns}^{-1}$, $\alpha = 5$, $\Delta\Phi_{rel} = 0$ in the upper panel of a) and $\Delta\Phi_{rel} = \pi$ in the lower panel of a). Panel b) illustrates the maximum cross correlation coefficient Σ_{corr} as function of the relative feedback phase $\Delta\Phi_{rel}$ for $p_{t,r} = 1.5$, $\kappa_c = 20\mathrm{ns}^{-1}$, and $\gamma_t = 20\mathrm{ns}^{-1}$. The figure depicts results for three different feedback strengths in the receiver ECSL: $\gamma_r = 10\mathrm{ns}^{-1}$ (black line), $\gamma_r = 20\mathrm{ns}^{-1}$ (gray line), and $\gamma_r = 30\mathrm{ns}^{-1}$ (dash-dotted line). Courtesy of J. Mulet.

coupling conditions, the cross-correlation smoothly degrades when $\Delta\Phi_{rel}$ is increased from zero, while it increases when $\Delta\Phi > \pi$. However, since the synchronization sensitively depends on the coupling conditions, which are of vectorial nature, we also expect that the ratio between the feedback strength and the coupling strength influences the synchronization properties.

To investigate the influence of changing the ratio between the feedback and the coupling strength on the synchronization properties, we increase the feedback rate to $\gamma_r = 20\,\mathrm{ns}^{-1}$ and $\gamma_r = 30\,\mathrm{ns}^{-1}$, respectively. For these conditions, we also calculate the cross-correlation behavior as function of $\Delta\Phi_{rel}$. The gray line presents the corresponding behavior for $\gamma_r = 20\,\mathrm{ns}^{-1}$, while the dash-dotted line shows the results for $\gamma_r = 30\,\mathrm{ns}^{-1}$. When we compare the three different feedback-coupling conditions for CC dynamics, we numerically find that the maximum cross-correlation for the optimal feedback phase condition is obtained for moderate coupling conditions and when both subsystems have the same feedback rate. Furthermore, a small deviation from the optimal phase condition $\Delta\Phi_{rel} = 0$ leads to a faster decrease in the correlation. This situation, in which the synchronization quality distinctly degrades even for small variations of $\Delta\Phi_{rel}$ offers potential for realization of a novel OOSK chaos communications system. We will address this interesting application of this specific synchronization property of the closed loop configuration in Chapter 5.5.1, in which we will discuss peculiar synchronization properties in terms of realization of practical encryption/decryption schemes.

In summary, we have experimentally and numerically demonstrated that excellent synchronization of coherence collapse dynamics can be realized in the closed loop configuration. In the long cavity regime, we were able to achieve robust general synchronization with high synchronization quality. For best conditions, we have measured cross-correlation coefficients between the dynamics of the transmitter and the receiver of up to 0.9. Analysis of the synchronization properties has revealed an interesting synchronization scenario in the LCR. In this scenario, the dynamics of the transmitter and the receiver system can be cyclically changed between well-synchronized states and almost uncorrelated states for varying the relative feedback phase. We have also disclosed that the ratio between the feedback and the coupling strength crucially influences the synchronization qualities of the systems. In this context, it is worth noting that we can also numerically find the solution for perfect synchronization, when we consider similar feedback strength in the transmitter and in the receiver system, but only weak coupling between the systems [196]. Nevertheless, the perfect synchronized condition is by far more sensitive, so that synchronization can be lost even for only small deviations between the parameters of the transmitter and the receiver system of the order of 1 % [198]. This high sensitivity of the perfect synchronization solution is probably the reason why we do not find this solutions in our experiments. The good synchronization quality which we can achieve, on the other hand, lets us expect that perfect synchronization can be principally realized if it is possible to find a slightly better matching pair of SLs. Nevertheless, the excellent properties of the general synchronization solution suggest that realization of the proposed chaos communication methods is feasible. In particular, the characteristic sensitivity of the synchronization on the feedback phase might also allow for realization of a novel OOSK encryption/decoding scheme.

These first results are very promising with respect to realization of a functional chaos communication system. However, we also know that best carrier signal properties can be found for short cavity dynamics. Therefore, we are also interested in whether it is possible to achieve suitable synchronization quality in the closed loop configuration for high-dimensional broadband SCR dynamics.

5.3.5 Synchronization in the Short Cavity Regime

In the previous section, we have seen that it is increasingly challenging to achieve good synchronization in the closed loop configuration for enhancing the dynamical bandwidth of the subsystems. Nevertheless, we are interested in synchronizing broadband chaotic carrier signals to realize good chaos communication schemes, which exhibit high bit rates and good security qualities. With regard to this, we need to investigate the possibility to synchronize high-dimensional broadband short cavity regime dynamics, as this dynamics offers superior signal properties to LCR dynamics. However, the demands on the corresponding experiments in the short cavity regime are extremely high. This is not only due to the high bandwidth of the dynamics, but even more because of the additional sensitivity of the dynamic of the subsystems on the feedback phase. Therefore, in the first experiments, we consider operation conditions for which we find SCR dynamics with only moderate bandwidth. For these conditions, we try to determine the operation conditions for which the dynamics of the transmitter and the receiver synchronize best. Based on these first results, in a second step, we slowly increase the bandwidth of the dynamics of the subsystems so that synchronization is maintained. In this way, we would like to achieve the challenging synchronization of broadband chaotic SCR dynamics.

Low Pump Parameter

In Chapter 3, we have seen that the bandwidth of the dynamics in the short cavity regime is only moderate when the SL is operated near threshold. We will use this regime to verify whether synchronization of SCR dynamics can be achieved in the closed loop configuration. For this task, we use the same experimental setup as for our studies in the long cavity regime. To guarantee operation of the ECSLs in the SCR, we drastically reduce the external cavity length of both, the transmitter and the receiver system to $L_{EC,t,r} = 3.7\,\text{cm}$. This length corresponds to a round trip frequency of $\nu_{EC,t,r} = 4.07\,\text{GHz}$. We set the pump parameter to $p = 1.09$ for which the relaxation oscillation frequency of the solitary HLP1400 SL corresponds to $\nu_{RO} = 1.5\,\text{GHz}$. Thus, the SCR requirement, $\nu_{RO} \leq \nu_{EC}$, is clearly fulfilled. The transmitter is subject to moderate feedback with a ratio of $r_t = 0.040$ for which we measure a threshold reduction of $\Delta r_t = -0.084$, so that we can adjust $\Delta\Phi_t$ for pronounced quasi-periodic pulse package intensity dynamics. For these conditions, we try to synchronize the dynamics of the receiver by optimizing, $\Delta\Phi_r$, r_r, and r_c, where r_c denotes the coupling ratio defined as the ratio between the power injected into the receiver and the power emitted by the transmitter.

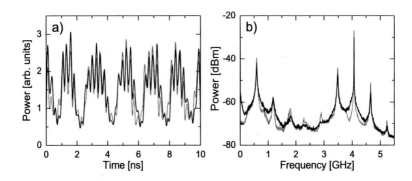

Fig. 5.9: Panel a) presents intensity dynamics of the transmitter (gray) and the receiver (black) for the close loop configuration in the short cavity regime for $p = 1.09$. The time series of the receiver system has been shifted forward in time by the coupling time $\tau_c = 6.6$ ns. The corresponding rf spectra are depicted in panel b). The rf spectra reveal the matching external cavity round trip frequencies of the subsystems located at $\nu_{EC} = 4.07$ GHz. Other conditions are: $L_{EC} = 3.7$ cm, $r_t = 0.040$, $r_r = 0.037$, and $r_c = 0.012$.

Figure 5.9 demonstrates that it is possible to achieve synchronization between the intensity dynamics of the transmitter and the receiver. We obtain best results when the receiver is also subject to moderated feedback of $r_r = 0.037$, for which we measure a threshold reduction of $\Delta r_r = -0.067$. The optimum coupling conditions are realized for $r_c = 0.012$ and for adjusted $\Delta\Phi_r$, such that $\Delta\Phi_{rel} \approx 0$. Hence, the optimal coupling and feedback conditions are similar to the conditions we have found for the excellent synchronization in the long cavity regime. The results for these conditions are presented in Figure 5.9. Panel a) shows intensity dynamics of the transmitter, depicted in gray, and of the receiver, represented in black. Both time series reveal quasi-periodic pulse package dynamics which is characteristic for the SCR. We find excellent agreement between the intensity dynamics of the transmitter and the receiver when we compensate for the delay of the receiver, being $\tau_c = 6.6$ ns. Again, this specific lagging of the receiver by the coupling time indicates that we have realized a generalized synchronization solution. In this particular example, we determine a cross-correlation coefficient of $\Sigma_{corr} = 0.76$. In addition to this, panel b) reveals excellent agreement between the rf spectra of the intensity dynamics of both subsystems. This combination of well-correlated temporal intensity dynamics and good agreement of the spectral properties of the dynamics highlights the quality of the synchronization.

However, as soon as we block the coupling between the transmitter and the receiver, the synchronization gets lost within about one nanosecond and the dynamics of both systems evolves independently.

This first demonstration of generalized synchronization of SCR dynamics in the closed loop configuration is very promising with respect to realization of a good chaos communication scheme. Nevertheless, we have found that the experiments are extremely challenging, due to the sensitivity of the transmitter and the receiver system on the feedback phase, which both are required to match. In addition to this, the experiments also give rise to high sensitivity of the synchronization properties on variations of the coupling conditions. The most crucial parameter in this configuration is the relative feedback phase, but we also find that small deviations of about 5 % from the optimum feedback strength lead to significant reduction of the achievable maximum cross-correlation coefficient. These first results already indicate that the higher bandwidth of the dynamics in the SCR has to be paid by smaller acceptable tolerances of the operation parameters and probably of the SL parameters, as well. Therefore, it is interesting also to study the resulting synchronization properties for higher bandwidth of the dynamics of the transmitter and the receiver system.

Moderate Pump Parameter

The optimized synchronization condition, which we have realized for $p = 1.09$, is the starting point for analysis of the synchronization properties of broadband SCR dynamics. From this condition, we increase the pump parameter of both SLs from $p = 1.09$ to $p = 1.26$ to enhance the bandwidth of the dynamics. For this pump parameter, we slightly optimize the remaining parameters to compensate for small deviations of the device parameters of the two SLs until we achieve best possible correlation between the dynamics of both subsystems.

The resulting dynamics are presented in Figure 5.10. In panel a) we depict a 10 ns long segment of the intensity dynamics of the transmitter (gray) and the receiver (black). For this level of pumping, the dynamics of both lasers does not give rise to the quasi-periodic dynamics anymore. Even more, we find very irregular broadband intensity dynamics. This property is verified by the corresponding rf spectra which are presented in panel b). The spectra of the dynamics reveal a bandwidth of about 4.5 GHz. In this range of frequencies the rf spectra are continuous and only vary within a range of about 15 dB. Up to this frequency, we find good agreement of the rf spectra indicating that both systems exhibit similar dynamical properties. However, if we compare the time series of the intensity dynamics and compensate for the delay of the receiver system, we find that the dynamics are only moderately correlated. For this

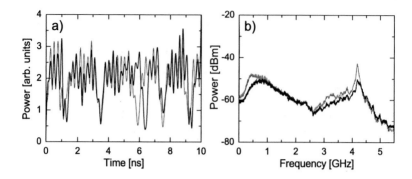

Fig. 5.10: Intensity dynamics of the transmitter (gray) and the receiver (black) for the close
loop configuration in the short cavity regime for $p = 1.26$. The time series of
the dynamics are shown in a), while the rf spectra are presented in b). Other
conditions agree to that of Figure 5.9.

state, we determine a cross-correlation coefficient of only $\Sigma_{corr} = 0.5$. A closer look
at the time series in panel a) reveals that the dynamics of the subsystems are highly
correlated in the interval between 0 ns and 4.5 ns. Then, suddenly, the amplitude of
the intensity pulsations of the receiver starts to deviate from that of the transmitter,
while the phase of the dynamics initially maintains synchrony. However, even the phase
relation between the pulsations gets lost at around 5.5 ns, until the receiver eventually
synchronizes to the transmitter again, at about 6.5 ns. It is remarkable that this
abrupt regain of synchronization directly leads to high correlation of the amplitude
and the phase of the intensity dynamics. Such intermittent loss of synchronization
can be frequently observed in the time series leading to the moderate time-averaged
cross-correlation coefficient.

We find that the synchronization properties do not improve by varying the operation conditions and the coupling parameters. Nevertheless, we can exclude mechanical
instability of the experimental setup as the origin of the continuously occurring intermittent loss of synchronization, since the corresponding nanoseconds time scale of the
loss and regain of synchronization is far beyond typical time scales of mechanical vibra-
tions which are of the order of several milliseconds. Therefore, we assume that these
intermittent losses represent characteristic synchronization properties which arise from
small differences of the dynamical properties of the coupled oscillators. These differences become increasingly important when the bandwidth of the dynamics enhances.
As a consequence thereof, in the following experiments, we decide to use a better

matching pair of SLs. With these SLs, we continue our studies of the synchronization properties of chaotic SCR dynamics. We focus on the influence of the relative feedback phase on the synchronization properties.

Influence of the Relative Cavity Phase

To gain further insight into the synchronization properties of ECSLs operating in the SCR in the closed loop configuration, we decide to use an even better matching pair of SLs. We note that finding such a pair of SLs is not a trivial task, since the parameters of device identical SLs can exhibit severe deviations of more than 10 %, even if the SLs are grown on the same waver. Nevertheless, we have found a well-matching pair of singlemode DFB SLs (IPAG[6]) which emit at a wavelength of 1539 nm. All but one of the accessible parameters of these SLs agree to better than 2 %. Only the slope efficiency of the SLs deviates by 4.5 %. These small deviations represent a significant improvement, as we will see in the following.

We adapt the experimental setup to the characteristics of the new SLs and try to achieve synchronization of broadband SCR dynamics. For the IPAG SLs, we find different feedback and coupling conditions for which we can achieve good synchronization if compared to the synchronization properties we have previously found for the HLP1400 SLs. This is due to the fact that the IPAG single mode DFB SLs are structurally different SL devices, so that they exhibit other dynamical properties. Thus, in the experiment, we observe best conditions when the transmitter and the receiver are subject to moderate feedback with $r_t = 0.060$ and $r_r = 0.038$, respectively. For these values, we measure a threshold reduction of $\Delta r_t = -0.165$ for the transmitter and $\Delta r_r = -0.140$ for the receiver SLs. The delay time of both subsystems amounts to $\tau_{EC,t,r} = 315\,\mathrm{ps}$, which corresponds to a cavity round trip frequency of $\nu_{EC,t,r} = 3.17\,\mathrm{GHz}$. Furthermore, we operate the SLs well-above threshold at $p_{t,r} = 1.3$ to achieve broadband chaotic emission dynamics. Under these conditions, we undirectionally couple the receiver to the transmitter system and optimize the coupling strength so that we can observe best synchronization. This is realized for $r_c = 0.041$. Finally, we readjust the feedback phases to achieve synchronization of chaotic broadband dynamics.

Figure 5.11 illustrates the synchronization properties for these conditions. The left column of the panels summarizes the results for synchronized conditions. The transmitter is always presented in gray, while the receiver is presented in black. The snapshot of the intensity dynamics in panel a) demonstrates well-synchronized intensity dynamics of the transmitter and the receiver system, comprising also the fast GHz pulsations of

---

[6]The characteristics of the IPAG single mode DFB SLs are summarized in Appendix A.1.

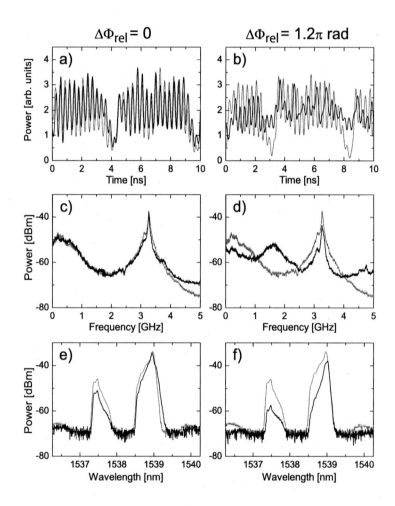

Fig. 5.11: Synchronization properties of the transmitter (gray) and the receiver (black) in
the closed loop configuration in the short cavity regime for two different values
for the feedback phase $\Delta\Phi_{rel} = 0$ and $\Delta\Phi_{rel} = 1.2\pi$ rad. Panel a) and b) depict
intensity time series. The corresponding rf and optical spectra are shown in
panels c) - d), and e) - f), respectively. The parameters are: $p_{t,r} = 1.3$, $r_t = 0.060$,
$r_r = 0.038$ $r_c = 0.041$.

the dynamics. We note that we also find general synchronization, indicated by the fact that the dynamics of the receiver lags the dynamics of the transmitter by the coupling time of $\tau_c = 5.3\,\mathrm{ns}$. In contrast to Figure 5.10, the dynamics does not give rise to intermittent loss of synchronization. Even more, in these first experiments, we find drastically improved correlation of the dynamics with cross-correlation coefficient of $\Sigma_{corr} = 0.8$, which still offers potential for further improvement by fine-tuning of the operation parameters. The good correlation is also reflected by the corresponding rf spectra of the intensity dynamics which are shown in panel c). Furthermore, the spectra reveal the characteristic broadband dynamics being typical for the SCR dynamics and which are desired for chaos communications. The results are completed by the optical spectra of the transmitter and the receiver illustrated in panel e). Due to the optical feedback, the sidemode suppression of the SLs decreases from 40 dB for the solitary SL to only 15 dB. Nevertheless, both the spectra are in good agreement in the complete spectral range indicating that the spectral dynamics are locked for synchronization.

This behavior drastically changes when we adjust for $\Delta\Phi_r = -1.2\pi$ shifting the relative feedback phase to $\Delta\Phi_{rel} = 1.2\pi$. This change results in loss of synchronization between the transmitter and the receiver dynamics. Now, the time series of the intensity dynamics in panel b) do not exhibit correlation of the intensity dynamics. Neither the amplitude nor the phase gives evidence for interdependence of the dynamics of both systems. The corresponding rf spectra are shown in panel d). The comparison of these spectra with the spectral properties for synchronization discloses drastic differences. Firstly, we find that the power of the receiver dynamics reduces. Secondly, we identify qualitatively different spectral behavior for the low frequency part and for the spectral region around the external cavity round trip frequency ν_{EC}. Thirdly, we identify a broad peak in the receiver spectrum at about 1.7 GHz. This peak first occurs at the low frequency edge of the spectrum for $\Delta\Phi_{rel} \approx 0.3\pi$ and shifts to higher frequencies for increasing $\Delta\Phi_{rel}$, until it vanishes near ν_{EC} for $\Delta\Phi_{rel} \approx 1.7\pi$. This scenario repeats cyclically with period 2π, highlighting its direct relation to the cyclic nature of the feedback phase. Interestingly, the optical spectra remain to be locked although the intensity dynamics does not give rise to correlation for $\Delta\Phi_{rel} = 1.2\pi$. However, the optical spectra in panel f) also show that the receiver emits less power due to the insufficient coupling to the transmitter dynamics.

These results reveal that the synchronization properties in the short cavity regime also sensitively depends on the relative feedback phase of the coupled subsystems. To elucidate this dependence, we provide an overview over the corresponding synchronization scenario by plotting Σ_{corr} as function of $\Delta\Phi_{rel}$ in Figure 5.12. The correlation discloses qualitative similar dependence as for the dynamics in the LCR, presented in Figure 5.7. Figure 5.12 displays high correlation coefficients for adjusted phase

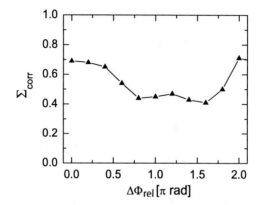

Fig. 5.12: Influence of the relative optical feedback phase $\Delta\Phi_{rel}$ on the cross correlation
coefficients of the intensity dynamics of the transmitter and the receiver in the
closed loop configuration. The system operates in the short cavity regime for
moderate pump parameters of $p_{t,r} = 1.3$.

$\Delta\Phi_{rel} \approx 0$, with well-agreeing rf and optical spectra. With increasing $\Delta\Phi_{rel}$, the correlation coefficient slowly decreases until it reaches its minimum in the interval around
$0.8\pi < \Delta\Phi_{rel} < 1.8\pi$. This interval of low correlation is relatively broad, if compared to
the results for the long cavity regime. Another difference is that the minimum cross-
correlation coefficient of about $\Sigma_{corr} \approx 0.4$ still indicates residual correlation of the
dynamics. In similarity with the behavior in the LCR, we also find a steeply increasing
Σ_{corr} for further incrementing $\Delta\Phi_{rel}$ to $\Delta\Phi_{rel} = 2\pi$, where synchronization is regained.

We have demonstrated that the synchronization quality can be improved by application
of better matching SLs. This result supports our hypothesis that the difficulties for
achieving synchronization of broadband SCR dynamics can be at least partly overcome
by reduction of the tolerable discrepancy between the transmitter and the receiver system. This also concerns the required precision of the operation parameters, since the
synchronization quality in the SCR very sensitively depends on the operation parameters. Fortunately, the operation parameters are well-accessible so that it is possible
to meet the high demands by well-designed experimental setups. Nevertheless, currently, time-consuming selection procedures have to be accepted to find well-matching
semiconductor lasers. However, these high demands on good agreement of the laser
parameters are beneficial to chaos communications, since an eavesdropper will not be
able to intercept the message by trying to synchronize a non-matching SL system. In

that sense, the allowable tolerances of the SL parameters define relevant security properties for the system. Therefore, better understanding of the influence of the internal SL parameters on the synchronization properties is desired. For this task, once more we take benefit of numerical modeling, since it is difficult to experimentally vary internal SL parameters.

5.3.6   Influence of Parameter Mismatch

We have seen that realization of good synchronization properties in the SCR regime requires very well-agreeing semiconductor lasers. However, usually it is difficult to find a pair of identical SLs. Therefore, it is essential to know which of the SL parameters are the most important with respect to the synchronization properties. From the experiments, we already have some experience concerning the requirements on some of these parameters which are well-accessible. The most relevant of these are the optical emission frequency, the threshold current, and the slope efficiency of the SLs. Nevertheless, other SL internal parameters also need to be considered. Since it is difficult to vary the internal SL parameters, we study their influence on the synchronization properties in the closed loop scheme numerically.

For this investigation, we use the established LK rate-equation model described by Eqs. 5.2 - 5.5 with the set of parameters presented in Table 5.4. We consider short external cavities with $L_{EC,t,r} = 3.9$ cm for which the delay time corresponds to $\tau_{EC,t,r} = 260$ ps. Both the transmitter and the receiver SL are subject to moderate feedback of $\gamma_t = 22.5$ ns^{-1} and $\gamma_r = 21.5$ ns^{-1}, respectively. For these conditions, and operation of the SLs at $p_{t,r} = 1.20$, the transmitter and the receiver system exhibit broadband chaotic intensity dynamics. In the modeling, we find high-quality general synchronization between the dynamics of both subsystems when we unidirectionally couple the SLs with a coupling strength of $\kappa_c = 12$ ns^{-1}. Then, we achieve high correlation of the intensity dynamics of $\Sigma_{corr} \approx 0.95$. This excellent synchronization quality serves as basis for our investigation of the influence of the system internal parameters on the synchronization properties.

Therefore, we numerically integrate the rate-equations Eqs. 5.2 - 5.5 for varying the internal parameters of the receiver SL and monitor the effect on the synchronization properties. We introduce the synchronization error which represents a good quantitative measure for characterization of the deviation from the optimum synchronization for which $\sigma = 0$.

$$\sigma(\Delta t) = \sqrt{\frac{\langle (P_t(t) - P_r(t - \Delta t))^2 \rangle}{\langle P_t^2(t) \rangle}} \tag{5.7}$$

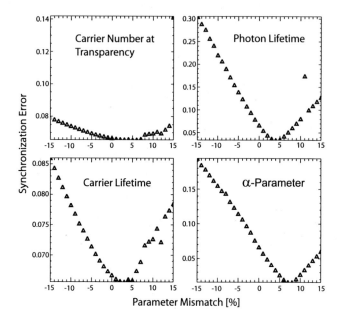

Fig. 5.13: Synchronization error for parameter mismatch in closed loop configuration in the
short cavity regime. The conditions are: $p_{t,r} = 1.2$, $\gamma_t = 22.5\,\text{ns}^{-1}$, $\gamma_r = 21.5\,\text{ns}^{-1}$,
and $\kappa_c = 12\,\text{ns}^{-1}$. Courtesy of R. Vicente.

As it can be seen from Eq. 5.7, the synchronization error σ can be easily calculated
from the intensity time series. We note, that for the corresponding calculations we
use $\Delta t = 0$, since the delay of the dynamics of the receiver has been rescaled. This is
justified by the observed generalized synchronization for which $\Delta \tau = \tau_c$.

Figure 5.13 shows the synchronization error versus parameter mismatch of the major
relevant internal SL parameters, i.e., the carrier number at transparency, the photon
lifetime, the carrier lifetime, and the α-parameter. We note that only one of the parameters is changed at a time, while the other parameters are assumed to agree perfectly.
Even in this case, the figure shows that a parameter mismatch of one of the system
parameters of the order of 10 % can result in severe degradation of synchronization
quality. Here the effect of each parameter is considered separately. In particular, we
find that the synchronization error most sensitively depends on the photon lifetime and
the α-parameter. Therefore, it is worth to pay special attention for the agreement of

Parameter	Closed Loop	
	Long Cavity	Short Cavity
$I_{th,sol}$	$< 8\%$	$< 6\%$
$\Delta P_{out}/\Delta I_{DC}$	$< 5\%$	$< 3\%$
p	$< 5\%$	$< 5\%$
$\Delta\omega_0$	-3..+5 GHz	-3..+5 GHz
	(-5..+10 GHz)	(-3..+5 GHz)
κ_c	-5 % +15 %	-5 % +10 %
	($\gamma > \gamma_{crit}$)	($\gamma > \gamma_{crit}$)
r	10 %	5 %
L_{EC}	$\pm 100\,\mu m$	$\pm 50\,nm$
$\Delta\Phi_{rel}$	$< \lambda/10$	$< \lambda/10$
α	$(\pm 10\%)$	$(+5..-10\%)$
γ_N	$(\pm 5\%)$	$(\pm 5\%)$
γ_p	$(\pm 5\%)$	$(\pm 5\%)$
robustness	less robust	sensitive

Table 5.5: Maximum tolerable parameter mismatch for synchronization in the closed loop configuration.

the α-parameter and the photon lifetime to achieve good synchronization.

These main results for the allowed tolerances of the mismatch of internal SL parameters are complemented by comprehensive experimental and numerical characterization of the acceptable mismatch of the remaining key parameters. These results are summarized in Table 5.5. Please note that the numerically obtained results are given in brackets to distinguish them from the experimental results. The results demonstrate that deviations of several percent are acceptable for mismatch of single parameters. However, simultaneous mismatch of many parameters will drastically reduce the synchronization quality. As a rule of thumb, we find that for experimental synchronization of SCR dynamics, the worst mismatch of the parameters should not exceed 5 %.

In summary, we have demonstrated that it is possible to achieve good chaos synchronization of broadband chaotic dynamics in the closed loop configuration, if the high demands on the allowed tolerances of the device parameters and on the accuracy of the operation parameters can be met. Under these premises, we find attractive synchronization properties which are promising for successful realization of functional chaos communication schemes. Nevertheless, our results also reveal that the synchronization in the closed loop scheme sensitively depends on the relative feedback phase. This means that a functional system which is based on this configuration needs to guarantee long-time stability of the external cavity alignment on sub-wavelength scale. Implementation of such a sophisticated system is technically feasible, but the required stability potentially causes additional costs. In this respect, we are also interested in the properties of simpler to be implemented and more robust configurations. Such an alternative is provided by the so-called open loop configuration. In the next section, we analyze the synchronization properties of SL systems in the open loop configuration and discuss whether it represents a suitable alternative to the closed loop configuration.

5.4   Synchronization Properties: The Open Loop Configuration

The principle of chaos communications is based on chaos synchronization phenomena. In general, the structure of a receiver system has to be matched to that of the transmitter to achieve chaos synchronization. Consequently, we have first investigated a receiver system that is structurally identical to the transmitter, leading to the closed loop configuration. Nevertheless, the strong dependence of the synchronization quality on the relative feedback phase highlighted the complexity of this "straight-forward" configuration. The complexity of the closed loop configuration might hamper functionality of practical applications. In the following, we therefore investigate the synchronization properties of a much simpler receiver structure, in which the receiver solely consists of a solitary SL. This configuration is referred to as the *Open Loop Configuration*.

5.4.1   Experimental Setup

In the open loop configuration, the receiver system consists of a solitary laser without feedback, while the transmitter consists of a SL with optical feedback. Figure 5.14 depicts the experimental setup of the open loop configuration. The SLs are the Hitachi HLP1400 Fabry-Perot SLs, we already used in previous experiments in the closed loop configuration. We also use the same control and detection devices for the experimental setup in the open loop configuration as for the closed loop configuration. Again, the two subsystems are optically coupled via injection of a well-defined fraction of the optical field of the transmitter system into the receiver laser. The double stage optical isolator (ISO) and the polarizer (POL) guarantee unidirectional coupling via the dominant TE

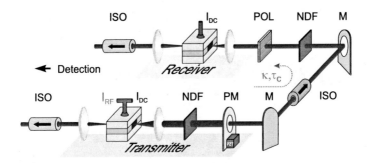

Fig. 5.14: Experimental setup of two unidirectionally coupled semiconductor laser systems in the open loop configuration.

component of the field. The strength of the coupling can be adjusted by placing a neutral density filter (NDF) between the transmitter and the receiver system. In the experiments, both lasers are operated well above the solitary laser threshold to allow for high dynamical bandwidth. The length of the external cavity of the transmitter can be varied so that synchronization in both dynamical regimes, in the LCR and the SCR regime, can be studied.

First, we try to achieve synchronization of fully developed chaotic LCR dynamics to find a good set of operation parameters. Then, we study whether synchronization can be also achieved for the attractive broadband dynamics in the SCR regime.

5.4.2 Synchronization in the Long Cavity Regime

The experiments in the long cavity regime focus on identification of good operation and synchronization parameters. For chaotic dynamics, this first has been done in the LCR, since the bandwidth of the dynamics is not as high as in the SCR regime. Furthermore, the dynamics in the LCR is insensitive against variations of the feedback phase, which simplifies the experiments. Thus, we choose the external cavity length to be $L_{EC,t} = 40$ cm, corresponding to $\tau_{EC,t} = 2.7$ ns and $\nu_{EC,t} = 375$ MHz. For this condition, we realize fully developed CC dynamics for the transmitter by adjusting for moderate feedback strength of $r_t = 0.050$ and for moderate pumping of $p_t = 1.15$. Then we couple the receiver, which is also pumped at $p_r = 1.14$ to the transmitter and optimize the coupling strength to achieve best possible synchronization. We find best synchronization between the receiver and the transmitter dynamics when the coupling

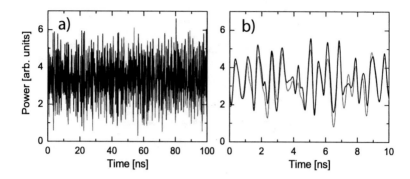

Fig. 5.15: Panel a): time series of intensity dynamics of the transmitter (gray) and the receiver system (black) in the coherence collapse regime in the open loop configuration for $p = 1.15$. Panel b) presents a zoom into the time series of a), highlighting the good synchronization. Other parameters are: $r_t = 0.050$, $r_c = 0.050$ and $\tau_c = 4.6$ ns.

strength is similar to the feedback strength in the transmitter system, $r_c = r_t = 0.050$.

The corresponding intensity time series of the transmitter (gray) and the receiver (black) are presented in Figure 5.15 a). We can identify that both SLs exhibit characteristic CC dynamics. Furthermore, the figure demonstrates that it is possible to achieve high-quality synchronization of LCR dynamics in the open loop configuration. We also find that the dynamics of the receiver lags by $\tau_c = 4.6$ ns giving evidence for general synchronization. It is worth noting that the CC dynamics of the receiver system is induced by the injection of light from the transmitter into the receiver laser. Without coupling, the receiver emits continuous wave, because it consists of a solitary SL without feedback. A 10 ns long zoom into the time series of panel a) is presented in panel b). The zoom discloses highly correlated intensity dynamics down to the fast time scale of the irregular GHz intensity pulsations. In this case, we determine a cross-correlation coefficient of $\Sigma_{corr} = 0.85$. This result underlines the excellent synchronization quality, which is also supported by the well-agreeing rf and optical spectra for these conditions. The synchronization is very robust against small perturbations and, in particular, against variations of $\Delta\Phi_t$. Furthermore, synchronization persists even for detuning between the optical frequencies of the two SLs of up to $\Delta\omega_0 \approx \pm 5$ GHz.

The observed high correlation coefficient of the dynamics and robustness of the synchronization indicate very attractive properties for realization of a practical chaos communication system. Therefore, we consider the open loop configuration as potential configuration and perform further experiments in which we comprehensively analyze its synchronization properties. With regard to possible applications, we concentrate on synchronization of chaotic SCR dynamics.

5.4.3 Synchronization in the Short Cavity Regime

To verify whether it is possible to realize suitable synchronization of broadband SCR dynamics in the open loop configuration, we reduce the length of the external cavity of the transmitter system to $L_{EC,t} = 3.9$ cm. This corresponds to a round trip frequency of $\nu_{EC,t} = 3.84$ GHz. Then, we slightly increase the pump parameter to $p_{t,r} = 1.2$ to allow for broadband chaotic intensity dynamics. For adjusted feedback phase of the transmitter system, the transmitter exhibits the characteristic chaotic SCR dynamics with a continuous rf spectrum that reveals high dynamical bandwidth. This is due to the fact that the transmitter ECSL operates well within the SCR, since the relaxation oscillation frequency amounts to $\nu_{RO} \approx 2.3$ GHz. In the next step, we optimize the coupling strength so that we achieve best possible synchronization between the dynamics of the receiver and the transmitter. We find good synchronization with highly correlated dynamics when the threshold reduction of the transmitter, due to the feedback, amounts to $\Delta r_t = -0.06$ and the threshold reduction for the receiver, induced by the injection from the transmitter, amounts to $\Delta r_r = -0.040$. These values correspond to $r_t \approx 0.06$ and $r_c \approx 0.05$.

We are able to synchronize the receiver to the chaotic SCR dynamics of the transmitter and find cross-correlation coefficients of up to $\Sigma_{corr} = 0.82$. Under these conditions, both rf spectra of the intensity dynamics of the lasers reflect the well-suited dynamical properties of the SCR which have been discussed in Chapter 3. The rf spectra demonstrate excellent agreement in a broad range of frequencies up to $\nu_{EC,t} = 3.84$ GHz demonstrating the good synchronization quality. To demonstrate the excellent synchronization of the dynamics even on sub-ns time scale, we present a 10 ns long segment of the corresponding emission dynamics in Figure 5.16. In the figure, the time series of the intensity dynamics of the transmitter is represented in gray, whereas the dynamics of the receiver is depicted in black. The time series of the receiver has been shifted forward in time by $\tau_c = 6.6$ ns to facilitate comparison.

The good synchronization properties suggest that the open loop scheme is ideally suited for realization of a practicable chaos communication scheme. However, we also find that it is increasingly challenging to achieve synchronization for enhancing bandwidth of the dynamics. This indicates that parameter mismatch becomes also important in the open loop configuration, although the tolerances seem to be less restrictive than for

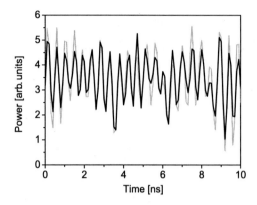

Fig. 5.16: Segments of chaotic SCR dynamics of the transmitter (gray) and the receiver
(black) for synchronization. The time series of the receiver has been shifted
forward by the coupling time of $\tau_c = 6.6$ ns.

the closed loop configuration. To verify the robustness of the open loop configuration, it
is essential to experimentally and numerically characterize the influence of parameter
mismatch on the synchronization quality. These studies are also required to define
acceptable tolerances for practical applications.

5.4.4   Influence of Parameter Mismatch

In this subsection, we analyze the robustness of the synchronization in the open loop
configuration. Therefore, we perform comprehensive experiments in which we detune
one of the system parameters until the synchronization quality degrades. To provide a
quantitative measure, we assume that synchronization is lost when the cross-correlation
drops below $\Sigma_{corr} < 0.7$. To complement the experimental results, we also apply numerical modeling to gain insight into the role of system-internal parameters which are
experimentally not accessible. The results are summarized in Table 5.6. The data
represent the maximum tolerances for which synchronization is maintained, under the
premise that only one parameter is detuned from the optimum conditions for synchronization. In the table, we distinguish between experimentally and numerically obtained
results, while the latter are enclosed by brackets. We also provide the corresponding
results for the closed loop configuration to allow for direct comparison of the synchronization properties of both configurations. The results reveal that the open loop

Parameter	Open Loop		Closed Loop	
	Long Cavity	*Short Cavity*	*Long Cavity*	*Short Cavity*
$I_{th,sol}$	$<8\%$	$<8\%$	$<8\%$	$<6\%$
$\Delta P_{out}/\Delta I_{DC}$	$<5\%$	$<5\%$	$<5\%$	$<3\%$
p	5-10%	$<5\%$	$<5\%$	$<5\%$
$\Delta\omega_0$	-5..+10 GHz	-5..+10 GHz	-3..+5 GHz	-3..+5 GHz
	(-5..+10 GHz)	(-3..+5 GHz)	(-5..+10 GHz)	(-3..+5 GHz)
κ_c	60%	60%	-5%+15 %	-5%+10 %
	$(\gamma > \gamma_{crit})$	$(\gamma > \gamma_{crit})$	$(\gamma > \gamma_{crit})$	$(\gamma > \gamma_{crit})$
r	not appl.	not appl.	10%	5%
L_{EC}	not appl.	not appl.	$\pm 100\,\mu$m	± 50 nm
$\Delta\Phi_{rel}$	not appl.	not appl.	$<\lambda/10$	$<\lambda/10$
α	$(\pm 20\%)$	$(\pm 20\%)$	$(\pm 10\%)$	$(+5..-10\%)$
γ_N	$(\pm 10\%)$	$(\pm 10\%)$	$(\pm 5\%)$	$(\pm 5\%)$
γ_p	$(\pm 10\%)$	$(\pm 10\%)$	$(\pm 5\%)$	$(\pm 5\%)$
robustness	very robust	robust	less robust	sensitive

Table 5.6: Comparison of the maximum tolerable parameter mismatch for synchronization in the open and the closed loop configuration.

configuration allows for similar mismatch of single parameters in the order of 5-10% as the closed loop configuration. Please note that the acceptable tolerances reduce when several parameters exhibit mismatch. Interestingly, the direct comparison of the results discloses that the synchronization of both configurations is fairly robust for LCR dynamics. Nevertheless, a significant difference consists in the sensitivity of the closed loop configuration on the relative feedback phase, while the open loop configuration is insensitive to $\Delta\Phi_t$ due to its different receiver structure. Comparison of the acceptable tolerances in the SCR, on the other hand, reveals that the open loop configuration is considerable less sensitive against mismatch of the SL parameters than the closed loop configuration. Therefore, we have been able to achieve good synchronization of SCR dynamics in the open loop configuration using the HLP1400 SLs. In contrast to this, we have not been successful to achieve good synchronization quality in the closed loop configuration using these SLs. We had to find better matching SLs to achieve comparable synchronization quality in the closed loop configuration. For that reason, the open loop configuration exhibits superior synchronization qualities with respect to realization of robust chaos communication schemes. However, to answer the question about the better configuration, it is essential to consider also the specific requirements of the applied encryption/decryption techniques on the synchronization quality. We will see that the performance and security properties of the applied encryption/decryption method need to be considered in addition for evaluation of the overall performance of the chaos communication system.

5.5   Message Encoding and Decoding

In this section, we investigate two different promising encryption/decryption methods. The first method is a chaos shift keying technique, while the second method represents a novel realization of ON/OFF shift keying that is based on the unique synchronization properties of the closed loop configuration. Both methods rely on chaos synchronization properties. In this context, we note that so far we have not provided rigorous evidence that the receiver dynamics is truly synchronized to the transmitter dynamics, instead of being only driven by the transmitter. Therefore, we need to find an experimental method that allows for distinguishing between both possibilities, since the functionality of chaos communication requires true synchronization.

This distinction between dynamical chaos synchronization and mere linear amplification can be made by verification of so-called chaos pass filter (CPF) properties [80]. For chaos synchronization, a small perturbation which eventually occurs in the transmitter signal is filtered out by the receiver which selectively synchronizes to the transmitter chaos only. This filtering process is linked to the existence of a stable synchronization manifold and is an intrinsically nonlinear process. Hence, if the configuration exhibits CPF properties, we can conclude that synchronization is stable. In contrast to this, a linear amplifier will amplify the chaotic signal and the perturbation in the same way.

Vice versa, a stable synchronization manifold implies CPF properties, which allows for discrimination of the perturbation and the chaotic signal. This mechanism can be applied for realization of CM, CMK, OOSK, and CSK in coupled ECSLs. Figure 5.17 illustrates the functionality of a chaos communication scheme which is based on chaos pass filter properties. In this chaos communication scheme, the message is applied in form of a small perturbation to the carrier signal. This perturbation can be implemented in different ways, e.g., as modulation of the pump parameter or the amplitude of the electromagnetic field in the cavity of the SL. Furthermore, the perturbation can be applied at different stages of the generation process of the carrier signal, e.g., inside the SL, inside the external cavity, or to the emitted electromagnetic field. Depend-

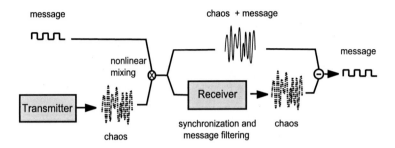

Fig. 5.17: Illustration of the functionality of a chaos pass filter and its utilization for encrypted chaos communication.

ing on the realized encryption/decryption technique, one distinguishes between the above-mentioned chaos communication schemes, which exhibit different performance properties. We consider a simple but efficient type of CSK, in which we introduce the message into the system in form of a small modulation of the pump parameter of the transmitter. The combined signal is sent to the synchronized receiver system. Since the synchronized receiver system exhibits CPF properties, it reproduces the chaotic carrier signal, but "damps" the superimposed perturbations, representing the message. Finally, the message can be recovered by comparison of the transmitted signal and the filtered signal.

To realize a functional chaos communication scheme, we need to investigate the synchronization properties in the closed loop and in the open loop configuration in terms of CPF properties to verify whether these configurations are suitable.

5.5.1 Chaos Shift Keying with Semiconductor Laser Systems

Chaos shift keying is one the most promising message encryption/decryption techniques, because it combines relevant advantages of chaotic carriers. On the one hand, the message can be nonlinearly mixed with the chaotic carrier signal, providing a dynamical key inherent to the transmission of the message. In this encryption procedure, the message spectrally spreads, which reduces the sensitivity of the system to narrow-band perturbations. On the other hand, the spectral spreading of the message complicates eavesdropping which is based on temporal and spectral filtering of the transmitted signal. For these reasons, the CSK technique provides higher security than chaos masking and chaos modulation keying methods. Consequently, we study this promising approach and verify whether CSK can be applied to synchronized chaotically emitting ECSLs. We are interested in a particular realization of CSK in which the message is encrypted via a small modulation of the pump parameter of the transmitter. This technique has two main advantages. Firstly, it is easy to implement. Secondly, it promises feasibility of high transmission rates. This is suggested by our results for the modulation response experiments in Section 5.2, which demonstrated that the maximum bandwidth for encryption is basically limited by the maximum bandwidth of the carrier signal. This implies that transmission rates in the multiple Gbit/s range seem to be feasible for chaotically emitting ECSLs, provided that the message can be successfully recovered at the receiver.

Chaos Pass Filter Properties of the Open Loop Scheme

In this subsection, we investigate the CPF properties of the open loop configuration. In similarity to the experiments on the modulation response, it is worth studying the synchronization properties in the long cavity regime for low pump parameters to better reveal the influence of characteristic system frequencies. Therefore, we set the pump parameter of both SLs to $p = 1.01$ and adjust the external cavity length of the transmitter system to $L_{EC,t} = 40 \, \text{cm}$, corresponding to a delay of $\tau_{EC,t,r} = 2.9 \, \text{ns}$. Furthermore, we apply moderate feedback of $r_t = 0.07$ and set the coupling strength to $r_c = 0.05$. For these conditions, we find good general synchronization with $\Sigma_{corr} = 0.75$ between the dynamics of the transmitter and the receiver, which lags by the coupling time of $\tau_c = 4.6 \, \text{ns}$.

To check for chaos pass filter properties in the present experiment, we impose a small sinusoidal current modulation upon the DC injection current of the transmitter, and observe the response of the receiver laser. The amplitude of the modulation corresponds to $\Delta p = 0.01$, while the frequency of modulation is $\nu_{mod} = 630 \, \text{MHz}$. Then we monitor the effect of the perturbation on the synchronized receiver dynamics. Figure 5.18 depicts the resulting rf spectra of the transmitter (gray) and the synchronized receiver (black). The modulation peak at $\nu_{mod} = 630 \, \text{MHz}$ is clearly visible in both spectra. We identify that the modulation peak in the receiver spectrum is damped by approximately 8 dB, whereas the chaotic broadband emission dynamics of the transmitter is well reproduced in the receiver system. Thus, the receiver selectively filters out the external perturbation evidencing a genuine dynamical synchronization process. The corresponding ratio between the peak power of the modulation in the transmitter and the corresponding peak power measured in the receiver provides a measure for the CPF properties. This ratio is referred to as message/carrier discrimination ratio (MCDR), since we intent to apply the message in the same way as the perturbation. For opti-

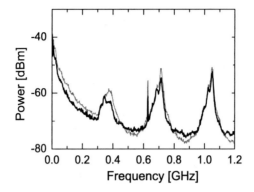

Fig. 5.18: rf spectra of the dynamics of the synchronized transmitter (gray) and receiver (black) for operation in the LFF regime for $p_{t,r} = 1.01$. The pump parameter of the transmitter is sinusoidally modulated at frequency of $\nu_{mod} = 630\,\text{MHz}$. The modulation amplitude is $\Delta p = 0.01$.

mized synchronization, we observe message/carrier discrimination ratios of up to 15 dB for modulation frequencies in the range between 50 MHz and 2 GHz. However, we also find that the message/carrier discrimination ratio sensitively depends on the modulation frequency. We find best discrimination for slightly lower frequencies than multiples of the external cavity round trip resonance $\nu_{EC,t}$, while the MCDR is about 3 dB less for slightly higher frequencies than these resonance frequencies. The discrimination ratio reduces to about 5 dB - 10 dB for modulation frequencies that exactly correspond to multiples of $\nu_{EC,t}$. We find worst discrimination for frequencies for which the power of the carrier signal is lowest.

For increasing the pump parameter, entering the CC dynamics regime, the open loop configuration gives also rise to sensitive dependence of the MCDR on the modulation frequency. Nevertheless, this dependence is less pronounced for CC dynamics, because of the smoother spectral distribution of the dynamics. Unfortunately, it turns out that the characteristic spectral dependence of the MCDR cannot be completely avoided in the LCR. This feature represents a serious drawback for realization of a practical chaos communication scheme, which is based on CSK. The reason for this is that a pronounced spectral dependence of the MCDR deteriorates the transmission of real messages, which are intrinsically broadband. Therefore, in the following, we focus on the CPF properties of synchronized SCR dynamics, since SCR dynamics offers better carrier signal properties.

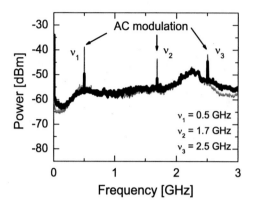

Fig. 5.19: Chaos pass filter properties in the open loop configuration in the short cavity
regime. The modulation amplitude of the pump parameter of the transmitter
amounts to $\Delta p = 0.03$

We reduce the external cavity length of the transmitter to $L_{EC,t} = 3.9\,\text{cm}$. This length
corresponds to a cavity round trip frequency of $\nu_{EC,t} = 3.84\,\text{GHz}$. Then, we adjust
for moderated feedback in the transmitter ECSL of $r_t \approx 0.06$ and realize moderate
coupling of $r_c \approx 0.05$. Finally, we set the pump parameter of both SLs to $p_{t,r} = 1.20$ to
achieve broadband chaotic SCR. For optimized conditions, we are able to realize good
synchronization between the dynamics of the transmitter and the receiver with $\Sigma_{corr} =
0.82$, for which we harmonically modulate the pump current of the transmitter with
an amplitude of $\Delta p = 0.03$. Again, we monitor the response of the receiver dynamics
in the rf spectrum from which we determine the MCDR. We repeat this procedure
for different modulation frequencies to determine the frequency characteristics of the
MCRD.

Figure 5.19 depicts the obtained rf spectra for three frequencies of modulation: $\nu_1 =
0.5\,\text{GHz}$, $\nu_2 = 1.3\,\text{GHz}$, and $\nu_3 = 2.5\,\text{GHz}$. The rf spectra of the transmitter dynamics
are presented in gray, while the rf spectra of the receiver dynamics are depicted in
black. Please note that the figure contains results of three subsequent measurements.
Figure 5.19 shows MCDRs between 3 dB and 12 dB depending on the frequency of modulation. We identify that the discrimination ratio in tendency decreases for increasing
modulation frequency. For continuously sweeping the spectral range of the dynamics,
we achieve good discrimination values of at least 10 dB in a broad range of frequencies
up to about 2.2 GHz with a maximum of 22 dB at about $\nu_{mod} \approx 0.3\,\text{GHz}$. The mea-

surements reveal that the discrimination reduces to an average of about 5 dB-10 dB for modulation frequencies beyond 2.5 GHz which is maintained up to $\nu_{EC,t} = 3.84$ GHz. For modulation frequencies exceeding $\nu_{EC,t} = 3.84$ GHz, the dynamics of the receiver gives rise to a drastic fall-off of the MCDR. From these results, we deduce the existence of a synchronization manifold for the open loop configuration with excellent stability properties.

Further experiments in the open loop configuration revealed that optimum conditions comprise MCDRs of about 10-15 dB in the full spectral range up to the cavity round trip frequency. The results also demonstrate that the frequency characteristics of the MCDR is smoother in the SCR than in the LCR. This combination of good MCDRs with smooth frequency characteristics is very attractive for realization of a functional chaos communication system. Nevertheless, also in the SCR we still find some variations in the frequency characteristics of the MCDR, even for small modulation amplitudes. Consequently, the question arises whether these characteristics of this nonlinear filter process indeed allow for successful transmission and recovery of a message, or if it significantly deteriorates the message recovery process. To anticipate this answer, we will later demonstrate that successful message transmission is possible in this configuration, but first we also analyze the CPF properties of the closed loop configuration.

Chaos Pass Filter Properties of the Closed Loop Scheme

In the previous section, we have seen that the open loop configuration reveals good CPF properties, from which we could conclude that the synchronization manifold in this configuration is stable. So far, we have identified better synchronization properties for the closed loop configurations. Therefore, we are interested in whether this observation is also reflected by the corresponding CPF properties. To allow for comparison with the results for the open loop scheme, we also study LFF dynamics in the long cavity regime. Therefore, we choose comparable operation parameters as for our experiments in the open loop configuration. We apply moderate feedback to both SLs and operate the SL near their solitary laser threshold currents at $p = 1.01$. The feedback ratio of the transmitter corresponds to $r_t = 0.07$, while we set the feedback ratio of the receiver to $r_r = 0.05$. Furthermore, we carefully match the length of the external cavities to $L_{EC,t,r} = 40$ cm, so that the delay amounts to $\tau_{EC,t,r} = 2.9$ ns. Finally, we couple the receiver to the transmitter with a strength of $r_c = 0.03$ and optimize the relative feedback phase to achieve best synchronization. We obtain excellent synchronization with cross-correlation coefficients of up to $\Sigma_{corr} = 0.9$. For these conditions, we apply a small harmonic current signal with an amplitude of $\Delta p = 0.01$ to the pump current of the transmitter SL to determine the CPF properties.

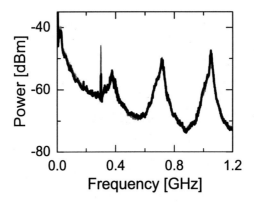

Fig. 5.20: rf spectra of the LFF dynamics of the synchronized transmitter (gray line) and
receiver (black line) for $p_{t,r} = 1.01$. The spectra show the response of the receiver
on a small sinusoidal modulation of the pump parameter of the transmitter. The
frequency of the modulation is $\nu_{mod} = 300\,\mathrm{MHz}$, and the modulation amplitude
corresponds to $\Delta p = 0.01$.

Figure 5.20 depicts the obtained rf spectra of the transmitter (gray), and the receiver
(black) for synchronized LFF operation. We can identify the modulation peak at
$\nu_{mod} = 300\,\mathrm{MHz}$, which is clearly visible in both spectra. However, the peak in the
receiver spectrum is strongly damped by more than $10\,\mathrm{dB}$, while the modulation does
not influence the excellent synchronization of the receiver to the chaotic carrier signal of
the transmitter. This good message/carrier discrimination ratio reveals robust stability
of the synchronization manifold. For optimized synchronization, we find MCDRs of up
to $20\,\mathrm{dB}$ for modulation frequencies between $50\,\mathrm{MHz}$ up to $2.2\,\mathrm{GHz}$. However, we also
find significant dependence of the MCDR on the modulation frequency with similar
frequency characteristics as for the open loop scheme. A distinct feature in the closed
loop configuration is that in contrast to the good synchronization quality for $\Delta\Phi_{rel} = 0$,
we do not observe CPF properties as soon as $\Delta\Phi_{rel}$ significantly deviates from zero,
such that a substantial reduction of the correlation between the time series can be
observed. In this case, we do not find the modulation signal in the dynamics of the
receiver. This gives evidence that the harmonic signal cannot be transmitted when
synchronization is lost.

The observed excellent CPF properties reflected by the high MCDRs indicate that
the closed loop configuration exhibits better synchronization quality than the open

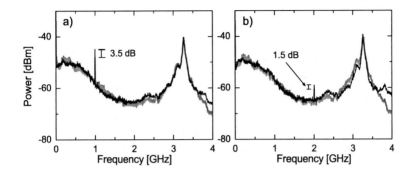

Fig. 5.21: rf spectra of the dynamics of the synchronized transmitter (gray lines) and receiver (black lines) for $p_{t,r} = 1.30$. The spectra show the response of the receiver on a small sinusoidal modulation of the pump parameter of the transmitter. The frequencies of the modulation are 1 GHz in panel a), and 2 GHz in panel b). The amplitude of the modulation is $\Delta p = 0.03$.

loop configuration. These good CPF properties offer great potential for practical applications in chaotic carrier communications, since they allow for clear discrimination between a message and the carrier signal so that a message can be reliably decoded by the receiver. However, our investigations also reveal the crucial importance of the optical feedback phase in the closed loop configuration. This combination of excellent CPF properties and sensitivity on the relative feedback phase opens up perspectives for a new chaotic communications scheme which we will present in the following subsection.

With respect to the conventional CM, CMK, and CSK methods, we are interested in the CPF properties of synchronized chaotic broadband dynamics, as can be realized in the SCR. For these experiments, we need to use the better matching pair of the singlemode IPAG SLs to achieve synchronization of the desired dynamics. The reason for this are the high requirements on the acceptable tolerances of parameter mismatch, which cannot be met with the HLP1400 SLs. After changing the SLs, we chose the same experimental conditions as in our previous experiment for which we were able to realize good synchronization quality. These conditions are: $p_{t,r} = 1.3$, $\nu_{EC,t,r} = 3.17\,\text{GHz}$, $r_t = 0.060$, $r_r = 0.038$, $r_c = 0.041$, and $\Delta\Phi_{rel} = 0$. Again, we are able to achieve fairly good synchronization of the dynamics of both subsystems with cross-correlation coefficients of up to $\Sigma_{corr} = 0.72$. Then, we add a harmonic current modulation with an amplitude of $\Delta p = 0.03$ to the pump current of the transmitter and monitor the response of the receiver the rf spectrum of its dynamics.

Figure 5.21 presents results for two different frequencies of modulation. In panel a) the modulation frequency corresponds to $\nu_{mod} = 1\,\text{GHz}$, while it has been shifted to $\nu_{mod} = 2\,\text{GHz}$ in panel b). Interestingly, the modulation experiment reveals quite low MCDR of less than 5 dB in the range between 50 MHz and 3.3 GHz, despite of the well-agreeing rf spectra of the transmitter and the receiver. These low CPF properties indicate that the synchronization is not as good as it might be, when only considering the spectral agreement of the intensity dynamics. In other words, the synchronization manifold is only weakly stable for these conditions. On the on hand, this observation underlines the relevance of measuring CPF properties to experimentally verify that the dynamics of the coupled systems indeed are synchronized. On the other hand, we can also see that cross-correlation coefficients of the order of $\Sigma_{corr} \approx 0.7$ are not sufficient for realization of a reliable chaos communications scheme. Therefore, the synchronization properties of the closed loop configuration in the short cavity regime need to be further improved. This probably can be achieved by better matching subsystems, which will be done in future work.

Nevertheless, the results indicate that we can achieve excellent synchronization quality for the closed loop configuration in the LCR, with better CPF properties than in the open loop configuration. This represents a relevant result with respect to realization of a chaos communication scheme utilizing LCR dynamics. However, the sensitivity of the closed loop configuration in the short cavity regime highlights once more that realization of such a chaos communication system is technically challenging, although it is principally possible. One motivation to take the additional technical efforts into account is motivated by the higher level of security, since a potential eavesdropper certainly has to cope with the enhanced difficulty to match his SL system to achieve the required quality of synchronization for intercepting the message. In that sense, a functional communication scheme needs to be tailored according to the demands on practicability, i.e., reliability and transmission bit rate, and security.

In the following, we benefit from the characteristic sensitivity of the synchronization quality of the closed loop configuration on the relative feedback phase, which can be used for realization of a fascinating communication scheme. The proposed scheme promises high levels of security and provides an attractive alternative to the known encryption/decryption methods.

5.5.2   The ON/OFF Phase Shift Keying Scheme

In this subsection, we propose a chaotic communication scheme which is based on the unique synchronization properties of the closed loop configuration. We refer to this scheme as the *ON/OFF Phase Shift Keying* (OOPSK), since its functionality principle is based on switching between synchronization and desynchronization. The physical basis for the functionality of the OOPSK is provided by the sensitivity of the synchronization in the closed loop configuration even to small variations of the relative feedback phase $\Delta\Phi_{rel}$. The idea is to rapidly change $\Delta\Phi_t$ to switch $\Delta\Phi_{rel} = \Delta\Phi_t - \Delta\Phi_r$, which in turn switches the receiver dynamics between synchronization and desynchronization. The advantage is that although the receiver reveals drastic changes in the dynamics, the phase change in the transmitter system does not give rise to different dynamics, neither in the rf spectrum, nor in the optical spectrum. Therefore, it is possible to encrypt a message by two different values of $\Delta\Phi_t$. By switching between these two phase values, we can encrypt a message. The message is decoded by the receiver which responds to the transmitted signal by switching between synchronization and desynchronization. Hence, these two states are used for binary encryption/decryption of messages. We assign bit "0" to synchronization and bit "1" to uncorrelated states. Thus, message recovery is simply accomplished by monitoring the synchronization error. Technically, it is possible to introduce such fast variations of $\Delta\Phi_t$ by inserting an electro-optical modulator into the optical feedback loop of the transmitter system. In the following we numerically demonstrate the feasibility of the OOPSK method. The presented numerical results have been obtained in collaboration with J. Mulet, P. Colet and C.R. Mirasso form the IMEDEA and the UIB, Palma de Mallorca, Spain.

Modeling Results ON/OFF Phase Shift Keying

Security aspects are important for encrypted chaos communication schemes. Sometimes, this issue is treated as a weak point in CM, CMK and CSK communication schemes, since in these methods often the message is encrypted by modulation of the amplitude of the carrier signal [78, 79, 172]. Therefore, many of the realized schemes require small amplitude of the message to prevent application of simple attacks using temporal or spectral filter techniques. This problem can be avoided with this newly developed OOPSK scheme, since the phase keying does note effect the spectral properties of the transmitter dynamics, while only a well-matched receiver can act as a sensitive detector for phase changes. Thus, the proposed OOPSK offers a potentially high level of security, since an eventual eavesdropper cannot decide whether or not a message is sent.

We numerically investigate the feasibility of the OOPSK by varying the feedback phase of the transmitter through a train of pseudo-random bits. For modeling, we use the LK rate-equations Eqs. 5.2 - 5.5 with the set of parameters presented in Table 5.4. We point out that this model has also correctly reproduced the synchronization scenario

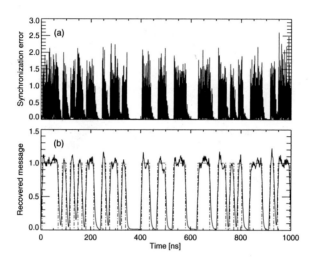

Fig. 5.22: ON/OFF phase shift keying scheme. The behavior of the synchronization error
as defined in Eq. 5.7 is depicted in panel a). The digital message at 64 Mbit/s
$\Delta\Phi_t(t)$ is represented by the dash-dotted line in panel b), while the recovered
message after filtering is presented by the solid line in panel b). The parameters
are: $\gamma_t = \gamma_r = 20\text{ns}^{-1}$, $\kappa_c = 30\text{ns}^{-1}$ and $p_t = p_r = 1.5$.

which we have experimentally uncovered for shifting $\Delta\Phi_{rel}$ between 0 and 2π. In
the modeling, we have varied $\Delta\Phi_{rel}$ by switching between $\Delta\Phi_t = 0$ (bit "0") and
$\Delta\Phi_t = 0.16\,\text{rad}$ (bit "1"), for which we find substantially reduced correlation between
the dynamics of the subsystems. To quantitatively determine the induced change in
the quality of correlation, we estimate the synchronization error of the normalized time
series as defined by Eq. 5.7. Figure 5.22 presents the synchronization error when a train
of pseudo-random bits at 64 Mbit/s is applied to $\Delta\Phi_t(t)$, which is now a function of
time. In Figure 5.22 a) we demonstrate that the synchronization error almost vanishes
for $\Delta\Phi_t = 0$, implying that $\Delta\Phi_{rel} = 0$. In contrast to this, the synchronization error
suddenly increases when the relative feedback phase is switched to $\Delta\Phi_{rel} = 0.16\,\text{rad}$.
We also find that the recovery of synchronization is slower, when switching back to
$\Delta\Phi_{rel} = 0$. These synchronization transients and, in particular, the recovery time
for synchronization, limits the achievable transmission rate of the OOPSK method.
For this set of parameters, we find that the maximum bit rate, which still allows for
message recovery, is of the order of 100 Mbit/s. To demonstrate that the message

can be recovered at the receiver, in Figure 5.22 b) we present the original message as dash-dotted line and the low-pass filtered synchronization error of panel a).

These synchronization transients limit the achievable transmission rates of the OOPSK scheme. Nevertheless, since the transients are directly related to the synchronization process in the closed loop configuration, further insight into the underlying mechanisms of the synchronization mechanism might disclose possibilities to speed up the synchronization process and, consequently, the feasible transmission rates. In spite of the restriction in speed the OOPSK method provides a considerable alternative to the previously known chaos communication schemes because it offers an increased level of security. We point out that previously known encryption/decryption methods, i.e., CM, CMK, and CSK also exhibit attractive security properties. The advantage of these methods consists in that they principally allow for high transmission rates in the order of multiple Gbit/s. Therefore, these systems represent very attractive alternatives for practical communication applications for which considerable amounts of private and confidential data need to be transmitted. In the following, we demonstrate that such a functional system can be realized by harnessing CPF properties.

5.5.3 A Functional Chaos Shift Keying Cryptosystem

Motivated by the successful synchronization of short cavity regime dynamics in the open loop configuration and in conjunction with its good CPF properties, we perform a digital signal transmission experiment for the described CSK encryption method via direct modulation of the pump parameter of the transmitter laser. The main advantage of utilizing this method in the chosen SCR consists in the fact that SCR dynamics offers continuous rf spectra, without conspicuous resonances which can deteriorate the efficiency of the message encryption. Nevertheless, the amplitude of the modulation has to be weak to prevent its fingerprints occurring in the rf spectrum. For encryption, optimum nonlinear mixing of the message into the dynamics is desired. Since we did not have a bit-pattern generator available, we met the requirements of more realistic modulation signals by using square wave (SQW) signals as test signals mimicking relevant features of binary messages. These signals allow for studying effects of the nonlinear encryption/decryption processes on the relevant steep slopes of digital messages. We note that an extension to random bit-pattern messages is required for a characterization of the system in terms of Bit-Error-Rate (BER) measures.[7] Nevertheless, the physics of the message encryption/decryption process can be studied well

---

[7]Although characterization of the system performance in terms of BER measures is very attractive, this project is beyond the scope of this thesis, which focused on investigation of the fundamentals of SL based chaos communications.

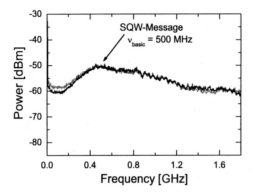

Fig. 5.23: rf spectra of the transmitter and the receiver laser for a square wave modulation
at a fundamental frequency of $\nu_{basic} = 500\,\text{MHz}$. The spectrum of the transmitter
is depicted in gray and that of the receiver in black.

by utilization of SQW-signals, providing deeper insight into the physical properties of
the communication system than the harmonic test signals.

In the corresponding experiment, we chose the amplitude of the SQW-signal such
that it cannot be recognized from the rf spectra. The corresponding rf spectra of the
transmitter and the receiver are depicted in Figure 5.23. The rf spectra show excellent
agreement, underlining the good synchronization quality of the configuration. The
fundamental modulation frequency ν_{basic} is located at 500 MHz, and the amplitude of
the modulation is $\Delta p = 0.03$. The main question is, if and how the digital signal, which
is nonlinearly mixed within the carrier, can be recovered after transmission. Demanding
processing should be avoided, since it would increase the complexity and the potential
costs of such a system significantly. A good compromise between simple realization
and good message/carrier discrimination needs to be found. Thus, we propose the
following procedure for message recovery:

1. Subtract the mean of the transmitter and the receiver output.

2. Normalize the variance of the receiver signal to the variance of the transmitter
   signal.

3. Subtract the normalized receiver signal from the normalized transmitter signal.

4. Filter the resulting signal with a multi-bandpass filter ($N \times 0.5\,\text{GHz} \pm 0.15\,\text{GHz}$).

5. Smooth the extracted message by low-pass filtering (2 GHz cut-off frequency).

We demonstrate that although no indications of the modulation can be noticed from the rf spectra, successful extraction of the signal is still possible. This is illustrated in Figure 5.24, which depicts 10 ns long segments of the intensity dynamics of the transmitter (upper gray line) and the receiver (upper black line). The original SQW-signal is presented as dashed gray line at the bottom of the figure. Additionally, the extracted signal, which has been obtained following the described procedure, is presented as black line, also at the bottom of the figure. The results clearly demonstrate that the message can indeed be extracted by application of this straightforward substraction process described above, and despite it was nonlinearly mixed in the optimized high-dimensional broadband carrier. We find that the extracted signal is somewhat distorted. We point out that these distortions originate from the nonlinear characteristics of the CPF properties and are not due to insufficient temporal averaging. Despite these distortions, such a binary signal can be easily processed by application of simple digital signal processing methods. Therefore, our results demonstrate that this tailored chaos communication scheme is fully functional. The system combines both, the beneficial properties of the optimized carrier which guarantee efficient encryption of digital messages at high bit-rates and the optimized synchronization properties which facilitate successful decryption of digital messages. We note that in our experiment the maximum modulation frequency is limited to 500 MHz by the pulse generator. Despite of these technical limitations, the carrier signal and the CPF properties suggest that an extension of the message frequency into the GHz range is possible. The fundamentally interesting question, in how far the distortions of the extracted signal are related to the synchronization process for the open loop configuration needs to be clarified in further experiments. The corresponding results are expected to allow for reduction of the message distortions which offers potential for further improvement of the efficiency of the encryption/decryption process.

Finally, we point out that it is not self-evident that this simple difference-signal extraction-process is a good choice in this configuration for the following reasons. For means of robustness, the utilized type of synchronization in our experiments is generalized synchronization. This type of synchronization implies that the dynamics of the transmitter and the receiver are not identical, even for best synchronization conditions. Based on this property, one might argue that these small deviations can result in decryption errors, which could impede the functionality of the real-world chaos communication. In particular, when additional disturbing effects such as dispersion effects and loss in optical fibers come in to play. Then, the encryption/decryption properties of the communication system needs to be quantified in terms of BER, while practical chaos communication systems have to fulfill specific telecommunication standards. Conse-

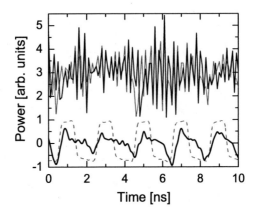

Fig. 5.24: Corresponding segments of the intensity time series of the transmitter and the
receiver laser. The time series of the transmitter is depicted in gray (upper line)
and that of the receiver in black (upper line). The original modulation SQW-signal
is represented in gray (bottom), while the extracted signal from the receiver is
depicted in black (bottom).

quently, it needs to be verified that transmission of real digital messages with multiple
Gbit/s rates is possible, even for distances exceeding several kilometers. This verification has been given in a field-experiment, in which we have been able to demonstrate
successful operation of a real-world chaos communication system that was implemented
into an existing fiber network [82]. In this first successful demonstration of a functional
real-world chaos communication system, we have used the slightly less sophisticated
chaos masking encryption/decryption method. Additionally, we have applied dispersion compensating fibers and optical amplifiers to compensate for dispersion and loss
in the transmission line. Implementing these actions, we were able to transmit an
encrypted pseudo-random digital message with a bit rate of 1 Gbit/s over a distance
of 120 km, while we achieved an BER $\approx 10^{-7}$ (without error correction procedures)
that is almost compatible with the requirements imposed by existing telecommunication standards. These promising results are expected to boost development of chaos
communications systems.

5.6 Summary and Discussion

In conclusion, we have investigated the possibility of utilizing the dynamical properties of chaotically emitting ECSLs for realization of functional chaos communications devices. Our studies have comprised three main aspects. Firstly, we have defined requirements on suitable carrier signals comprising performance and security aspects. Then, we have analyzed the dynamical properties of chaotically emitting ECSLs to find operation conditions that allow for generation of dynamics which meet the demands of chaotic carrier signals. We have identified two suitable regimes of chaotic dynamics. These regimes comprise fully developed coherence collapse dynamics in the long cavity regime and broadband chaotic dynamics in the short cavity regime. Both dynamics can be realized by application of moderate feedback and moderate pump parameters. Secondly, we have studied the possibility to synchronize a receiver system to the chaotic carrier signals generated by the transmitter system. Since the principles of chaos communications are based on harnessing chaos synchronization phenomena, the synchronization properties of a functional system need to fulfill specific requirements. We have considered two promising configurations. In the closed loop configuration, the receiver system is similar to the transmitter system and consists also of an ECSL. In contrast to this, the receiver only consists of a solitary SL in the open loop configuration. We demonstrated that chaos synchronization can be achieved in both configurations. Therefore, we have comprehensively characterized the particular synchronization properties. Our results revealed that both configurations can exhibit the required quality of synchronization, allowing for encryption and decryption of messages. It turned out that the more robust open loop configuration exhibits slightly less good synchronization quality than the closed loop configuration. The closed loop configuration, in turn, reveals higher sensitivity of the synchronization quality than the open loop configuration. We have demonstrated that even small parameter mismatch between the transmitter and the receiver system can drastically reduce the synchronization quality in the closed loop configuration. Thirdly, we have addressed the question of good encryption and decryption methods which allow for successful transmission of messages. Considering the requirements of practical applications on transmission rate and security, we have proposed two different encryption/decryption schemes, the ON/OFF phase shift keying and the chaos shift keying. The ON/OFF phase shift keying (OOPSK) is based on the sensitive dependence of the synchronization quality on the relative feedback phase which can be observed in the closed loop configuration. We have found that the advantage of the OOPSK method consists in its potentially high level of security, while the maximum achievable bandwidth for this method is restricted to about 100 Mbit/s, limited by the synchronization time of the receiver. In the proposed CSK method the message is encrypted via modulation of the pump parameter of the transmitter. The

Properties	Open Loop		Closed Loop	
	Long Cavity	*Short* Cavity	*Long* Cavity	*Short* Cavity
param. control	very good	very good	good	sensitive
robustness	very good	very good	sensitive	very sens.
compactness	no	yes	no	yes
white rf spectr.	no	yes	no	yes
mod. index	up to 15 dB	up to 20 dB	up to 15 dB	up to 20 dB
corr. time	$\gg 1\,\mathrm{ns}$	$< 1\,\mathrm{ns}$	$\gg 1\,\mathrm{ns}$	$< 1\,\mathrm{ns}$
attractor dim.	≈ 100	≈ 10	≈ 100	≈ 10
$d_{KY}(\sharp_{\lambda_i > 0})$	$\gg 10$	~ 10	$\gg 10$	~ 10
λ_{max}	$\lambda_{max}(L_{EC})$	max. in SCR	$\lambda_{max}(L_{EC})$	max. in SCR
$h_{KS}(\Sigma_{\lambda_i > 0})$	saturates	$h_{KS,max}$	saturates	$h_{KS,max}$
C.P. Filter	good	good	excellent	good but sens.
discrimination	up to 15 dB	up to 20 dB	up to 20 dB	up to 25 dB
min. achievable bandwidth	2 GHz	2.7 GHz	2 GHz	2.7 GHz
encryption	CM, CMD CSK	CM, CMD CSK	CM, CMD CSK, OOPSK	CM, CMD CSK

Table 5.7: Performance of open and closed loop chaos communication schemes.

advantage of this method is that it is easy to implement and that the message is non-linearly mixed into the carrier signal. This allows for realization of transmission rates of several Gbit/s, while providing good levels of security. Finally, we have coalesced all these results and demonstrated a functional chaos communication system which allowed for successful transmission and recovery of a 500 MHz test signal, applying the CSK method.

These presented studies verify that chaotically emitting SL systems are well-suited for realization of future communication schemes that allow for transmission of private and confidential data. Our results further show that the technical requirements increase when aiming for higher transmission rates. Therefore, it is promising to implement more sophisticated devices and technology. For the presented schemes it is possible to optimize the performance of the chaos communication system according to the specific requirements of the application. To provide a basis for future work and to allow for such an optimization, we summarize the key results of our studies in Table 5.7. The table provides a detailed overview over the characteristics of the different promising configurations, allowing for direct comparison of the proposed alternative systems.

5.6.1 Comparison of Open and Closed Loop Performance

The results in Table 5.7 comprise both, performance properties and security aspects, which need to be considered when discussing the question of the best configuration.

Robustness and Reliability

If we compare the technical properties, i.e., parameter control, robustness and compactness of the system, the robust open loop configuration seems to trump the closed loop configuration. Since the spectral properties of the carrier signal are also relevant for realization of reliable communication systems with high transmission rates, broadband carrier signals with high modulation index are desired, which can be realized for operation of both configurations in the short cavity regime. Hence, the performance of the open loop configuration with dynamics in the short cavity regime is favored. However, since we are interested in encrypted communications, the security aspects need to enter these considerations.

Security Aspects

The security of a good chaos communication scheme essentially depends on three factors: the dynamical properties of the carrier signal, the synchronization properties of the specific configuration, and the applied encryption/decryption method. For high levels of security, it is desired that the spectral properties of the carrier are similar to that of white noise. Therefore, chaotic dynamics comprising high spectral bandwidth with smooth spectral distribution is preferred. Additionally, high-dimensional (d_{KY}) dynamics with fast decreasing correlations and high entropy (h_{KS}) is desired. These requirements can best be fulfilled in the short cavity regime, still favoring the open loop configuration. However, when we consider the security properties which are defined by the synchronization properties, the closed loop configuration outperforms the open loop configuration. This is due to the fact that the fundamental mechanism for chaos communications is based on CPF properties. The quality of these properties are essentially defined by the stability of the synchronization manifold. Since the synchronization in the closed loop configuration reveals higher sensitivity of the CPF properties on small mismatch of the system parameters, it is more difficult to achieve synchronization without knowledge about the system and the operation parameters. In that sense, the number of relevant parameters and the sensitivity of the synchronization on parameter detuning crucially influences the level of security of the system. Or in other words, the number of dynamically relevant system parameters and the sensitivity of the synchronization on these parameters define the length of the cryptographic key that needs to be shared between the transmitter and the receiver. A potential eavesdropper who does not have information about the relevant system parameters has two options to crack the system. Either he finds a way to determine the hardware and the operation parameters for physically accessing the message, or he succeeds in estimating the dynamical parameters, which is also difficult because of the high-dimensional dynamics with large entropy. Under consideration of these aspects, the open loop configuration still reveals good properties, but the closed loop configuration promises even higher levels of security. Finally, the applied encryption/decryption also enters the security properties in that it is relevant how good the message is encrypted within the carrier. The favored methods are CSK and OOPSK.

Balancing all these aspects, we can state that with respect to high transmission rates and good levels of security both, the open loop and the closed loop configuration, offer best performance for short cavity regime dynamics and application of CSK. However, when easy-to-implement systems are desired, the open loop configuration is favored. On the other hand, when secure transmission is in first place, application of OOPSK in the closed loop scheme in the long cavity regime offers a good alternative.

5.6.2 Conclusion and Outlook

In conclusion, the presented results elucidate the high potential of chaotically emitting ECSLs for studying synchronization phenomena of high-dimensional broadband chaotic dynamics. We have demonstrated that coupled ECSLs can exhibit interesting synchronization properties, which, in turn, can be utilized for realization of novel applications. In particular, we have demonstrated the first fully functional chaos communication system which allows for encrypted transmission of data at transmission rates in the Gbit/s range while also providing high levels of security.

Nevertheless, the results also give rise to open questions and problems that deserve further investigation. The main problem of chaos communication systems consists in the difficulty of defining a measure for the level of security. Such a measure needs to be defined to allow for comparison with complementary cryptographic schemes. At the moment, it is unclear how the various aspects of security can be considered in one general measure. The largest difficulty arises from the fact that synchronization properties essentially define the level of security. Since synchronization properties usually reflect characteristics of the compound (coupled) system, it is problematic to define security in terms of the properties of a specific synchronization manifold of a particular receiver structure, since different receivers might lead to different results. To answer this question, further insight into the fundamental mechanisms leading to the various synchronization phenomena is essential. Such insight might be obtained from the novel concept of consistency, which has been proposed by Uchida et al. [202]. In this concept the response of a nonlinear dynamical system on the same driving signal is studied in consecutive experiments. Since the dynamical state of the receiver is different each time the same signal is injected, the response of the receiver can provide insight into fundamental synchronization properties.

We expect that these studies will provide results that might solve the question about how the level of security of chaos communication systems can be rigorously described. This concept of consistency might provide results which allow for realization of more sophisticated second-generation chaos communication schemes which are based on other types of synchronization than identical or general synchronization. Research on chaos communication systems has just begun. At this early days of chaos communications, we cannot exclude that less obvious forms of synchronization, such as phase synchronization, might allow for systems with superior security properties.

Summary

In this thesis, we have investigated the close connection between Nonlinear Dynamics (NLD) and Semiconductor Laser (SL) physics by experimentally studying the emission properties of SLs with delayed optical feedback. Our investigations have focused on the following two fundamental problems:

- High-dimensional temporally and spectrally chaotic dynamics of semiconductor lasers with optical feedback,

- Synchronization of broadband high-dimensional chaotic dynamics.

The investigations of high-dimensional chaotic dynamics and synchronization phenomena were motivated by two reasons. On the one hand, fundamental insight into dynamical phenomena of high-dimensional chaotic dynamics was desired. On the other hand, understanding and utilization of the nonlinear dynamical properties of SLs promised high potential for realization of novel applications. We have pursued this concept by studying two attractive examples:

- Tailoring the emission properties of semiconductor lasers by controlling their nonlinear dynamics properties,

- Encrypted communications based on synchronized chaotic carrier signals.

In the first integral part of this thesis, we have experimentally investigated the emission dynamics of chaotically emitting SLs with time delayed feedback. These systems are also referred to as *External Cavity SLs* (ECSLs). Our investigations focused on high-dimensional broadband chaotic dynamics. For characterization and analysis, we have applied state-of-the-art measurement technology and methods from NLD to gain first deep insight into the fundamental dynamical phenomena within this regime. We have demonstrated that the dynamics sensitively depends on the relevant systems parameters, i.e., the pump current, the feedback strength, the feedback phase, and the delay time of the optical feedback. The application of complementary modeling proved to be fruitful, since it allowed for utilization of NLD methodology for characterization of the dynamical properties which were beyond experimental accessibility. Our results have demonstrated the potential of ECSLs for generation of pronounced chaotic intensity dynamics with typical attractor dimension of the order of 10 and substantially greater, and high dynamical bandwidth beyond 10 GHz. Even more, we have revealed a close connection between the intensity and the spectral dynamics which, indeed, can be utilized for tailoring the coherence properties of SLs via influencing the dynamical properties of the system. In this context, the high versatility of the dynamics of ECSLs explicitly offers potential for implementation of SLs in new fields of applications.

In the second integral part of this thesis, we have applied the obtained knowledge about the dynamical properties of ECSLs to design a SL system with tunable dynamical emission properties. Our aim was to drastically reduce the coherence length of a SL, since coherence effects often limit the resolution of metrology applications which are based on SLs. Therefore, we have chosen an optimized SL and tailored the feedback conditions to enhance the coupling between the longitudinal Laser Modes (LMs). Following this approach, we have succeeded in realizing a SL light source in which the coherence length can be drastically reduced by inducing broadband chaotic emission of more than 130 LMs. The system offers a wide tuning range of the coherence length between 8 m and 130 μm by variation of the optical phase of the feedback. In this tuning scenario, the dynamics changes from stable emission of one (single) LM to chaotic emission of more than 130 LMs, comprising an optical bandwidth of about 7 nm. The achieved minimal coherence length is very attractive for applications in which bright incoherent light sources are required, such as for modern chaotic light detection and ranging, or for coherence tomography. This particular example demonstrates that NLD can indeed be successfully applied for tailoring the emission properties of SLs, while the designed system can also serve as a model-system for studying fundamental dynamical properties of pronounced multimode dynamics.

In the third integral part of this thesis, we have concentrated on synchronization phenomena of unidirectionally coupled SL systems. We were interested in the question whether it was possible to synchronize high-dimensional broadband dynamics of two coupled SL systems. In addition to our motivation from the fundamental NLD point of view, we were particularly interested in whether it was possible to harness synchronization properties of coupled SL systems for the realization of a functional *Chaos Communication System* which allows for encrypted communication. Functionality of chaos communication systems is based on particular properties of chaotic carrier signals, robust synchronization of such carriers, and appropriate methods for efficient, nonlinear encryption/decryption of messages. Therefore, our goal was to verify whether the dynamical properties of coupled ECLSs can fulfill all the requirements to allow for realization of a functional chaos communication system. In this context, we could benefit from the our insight into the dynamics properties of ECSLs obtained in the first part of this thesis. We have demonstrated that the operation conditions of ECSLs can be optimized to generate well-suited broadband high-dimensional chaotic carrier signals allowing for efficient encryption of a message. Subsequently, we have verified that excellent *Chaos Synchronization* with highly correlated dynamics can be achieved for such optimized chaotic carrier signals. This was possible in two different configurations of receiver systems, the *Open Loop* and the *Closed Loop* configuration. We have characterized the synchronization properties of both configurations to gain insight into the stability properties of the synchronization. We have given first experimental evidence for the existence of a stable synchronization manifold for such high-dimensional broadband chaotic dynamics. We have revealed excellent chaos pass filter properties (CPF) which manifest themselves in that the receiver can discriminate between the transmitted chaotic carrier signal and messages, which can be applied in form of small perturbations of the carrier signal. We have investigated the synchronization and CPF properties of both configurations and have evaluated the properties in terms of security aspects and practicability for chaos communications. Based on our obtained results, we have investigated to alternative chaos communication schemes, the newly developed *ON/OFF Phase Shift Keying* scheme which offers high levels of security, and a tailored *Chaos Shift Keying* scheme, for which we have realized a fully functional chaos communication system. For this system we have demonstrated successful transmission of a digital signal. The realized chaos communication system offers good levels of security and high transmission rates of up to several Gbit/s. Finally, we have compared the characteristics of different configurations providing a basis for continuative work on this fruitful application of chaos synchronization.

In conclusion, the presented results have accentuated the close connection between the scientific fields of NLD and SL physics. On the one hand, we have given evidence that SL systems can serve as excellent well-controllable model-systems for experimental NLD. On the other hand, we have successfully utilized NLD for controlling the dynamical properties of SL systems, which, in turn, demonstrated that applied NLD can open up practical applications for SLs. Our results have demonstrated that this beneficial link between NLD and SL physics can serve as a fruitful basis for future progress in both scientific fields.

Zusammenfassung

In dieser Arbeit wurde die enge Verbindung zwischen den beiden wissenschaftlichen Gebieten der Nichtlinearen Dynamik (NLD) und der Halbleiterlaserphysik untersucht. Dazu wurden die Emissionseigenschaften von Halbleiterlasern (HLL) mit verzögerter optischer Rückkopplung experimentell untersucht, wobei der Schwerpunkt auf der Erforschung folgender grundlegender Phänomene lag:

- Hochdimensionale, zeitliche und spektrale chaotische Dynamik von Halbleiterlasern mit optischer Rückkopplung,

- Synchronisation breitbandiger, hochdimensionaler chaotischer Dynamik.

Die Erforschung von hochdimensionaler chaotischer Dynamik und von Synchronisationsphänomenen war doppelt motiviert. Einerseits wurde aus Sichtweise der NLD ein grundlegender Einblick in die dynamischen Phänomene hochdimensional chaotischer Dynamik angestrebt, andererseits verspricht ein solcher Einblick eine mögliche Nutzbarmachung der nichtlinear-dynamischen Eigenschaften von HLL mit einem hohen Potenzial zur Realisierung neuartiger Anwendungen. Dieses Konzept wurde durch Untersuchung von zwei vielversprechenden Beispielen verfolgt:

- Maßschneiderung der Emissionseigenschaften von HLL durch Kontrolle der dynamischen Eigenschaften,

- Verschlüsselte Datenkommunikation basierend auf synchronisierten chaotischen Trägersignalen.

Im ersten Hauptteil dieser Arbeit wurde die Emissionsdynamik chaotisch emittierender HLL mit optischer Rückkopplung untersucht. Besonderes Interesse galt der Untersuchung hochdimensionaler chaotischer Dynamik mit hoher Bandbreite. Zur Charakterisierung und Analyse der dynamischen Eigenschaften dieser HLL-systeme wurden sowohl moderne Messtechnik als auch Methoden der NLD eingesetzt, um erste Einblicke in die zugrunde liegenden dynamischen Phänomene in diesem Regime zu erhalten. Es stellte sich heraus, dass die Dynamik wesentlich durch folgende Systemparameter bestimmt wird: dem Pumpstrom, der Rückkopplungsstärke, der Rückkopplungsphase und der Verzögerungszeit der Rückkopplung. Komplementär zu den Experimenten wurden Modellierungen durchgeführt, die Anwendung der Methodik der NLD zur Charakterisierung der dynamischen Eigenschaften ermöglichten, die experimentell nicht zugänglich waren. Die Resultate zeigten, dass HLL mit optischer Rückkopplung ein hohes Potenzial zur Erzeugung ausgeprägter chaotischer Intensitätsdynamik mit typischen Attraktordimensionen in der Größenordnung von 10 und deutlich höher, sowie hoher dynamischer Bandbreite jenseits von 10 GHz aufweisen. Insbesondere konnte eine enge Beziehung zwischen der Intensitätsdynamik und der Spektraldynamik aufgezeigt werden, die eine kontrollierte Manipulation der Kohärenzeigenschaften von HLL durch Beeinflussung der Dynamik ermöglicht. Diese Möglichkeit zur Maßschneiderung der Emissionseigenschaften eröffnet ein hohes Potenzial zur Erschließung neuer Anwendungsgebiete für HLL.

Im zweiten Hauptteil dieser Arbeit wurde das gewonnene Wissen über die dynamischen Eigenschaften dieser HLL-Systeme angewandt, um ein System mit abstimmbaren Emissionseigenschaften zu entwickeln. Ziel war es, die Kohärenzlänge eines HLL drastisch zu reduzieren. Dies war durch die Tatsache motiviert, dass Kohärenzeffekte oft die Auflösung von optischen Entfernungsmessgeräten einschränken. Zur Realisierung dieses Vorhabens wurde ein langer HLL ausgewählt, der optimierter Rückkopplung ausgesetzt wurde. Dadurch konnte eine effiziente Kopplung zwischen den longitudinalen Lasermoden (LM) erreicht werden. Diese Strategie ermöglichte eine drastische Reduzierung der Kohärenzlänge des HLL, die sich durch optische Breitbandemission mit mehr als 130 beteiligten LM auszeichnet. Das realisierte System bietet einen weiten Durchstimmbereich der Kohärenzlänge zwischen 8 m und 130 μm, der durch Variation der optischen Rückkopplungsphase zugänglich ist. Bei Durchstimmen dieses Kohärenzlängenbereiches wird die Dynamik zwischen stabiler Emission auf einer LM bis hin zu chaotischer Emission von mehr als 130 beteiligten LM, die eine optische Bandbreite von etwa 7 nm umfasst, geändert. Die erreichte minimale Kohärenzlänge von 130 μm ist attraktiv für Anwendungen, in denen leistungsstarke, inkohärente Lichtquellen benötigt werden, wie z.B. zur Entfernungsmessung mittels chaotischem Licht oder zur Kohärenztomografie. Somit konnte anhand dieses Beispiels gezeigt werden, dass sich die NLD erfolgreich zur Maßschneiderung der Emissionseigenschaften von HLL anwenden lässt, wobei das entwickelte System auch dazu dienen kann, grundlegende Eigenschaften ausgeprägter Multimodendynamik zu untersuchen.

Der dritte Hauptteil dieser Arbeit war der Untersuchung von Synchronisationsphänomenen unidirektional gekoppelter HL-Systeme gewidmet. Das Hauptinteresse galt der Untersuchung der Fragestellung, ob es möglich sei, breitbandige, hochdimensionale chaotische Dynamik zu synchronisieren. Zusätzlich zu dem fundamentalen Interesse an dieser Fragestellung, wurde auch die Möglichkeit der Nutzung von Synchronisationseigenschaften zur Realisierung funktionstauglicher *Chaos-Kommunikationssysteme* erörtert. Die Funktionalität effizienter *Chaos-Kommunikationsschemata* basiert auf möglichst komplexen, breitbandigen, chaotischen Trägersignalen, robuster Synchronisation dieser Trägersignale, sowie nichtlinearer Ver- und Entschlüsselung der zu sendenden Nachrichten. Ziel war es herauszufinden, ob die dynamischen Eigenschaften gekoppelter HLL mit optischer Rückkopplung diesen Anforderungen genügen können, und somit die Realisierung funktionaler Chaos-Kommunikationssysteme ermöglichen. Auch in diesem Zusammenhang konnte von dem gewonnen Einblick in die dynamischen Eigenschaften dieser HLL-Systeme profitiert werden. Dies ermöglichte die Erzeugung gut geeigneter breitbandiger, hochdimensionaler chaotischer Trägersignale, die eine effiziente Verschlüsselung von Nachrichten erlaubt. Weiterhin konnte erstmals nachgewiesen werden, dass hervorragende *Chaos-Synchronisation* mit hochkorrelierter Dynamik derartig optimierter Signale realisierbar ist. Dies wurde für zwei unterschiedliche Konfigurationen des Empfängersystems demonstriert, in der so genannten *Open Loop* und in der *Closed Loop* Konfiguration. Um einen tieferen Einblick in die Stabilität der Synchronisation zu erhalten, wurden die Synchronisationseigenschaften beider Konfigurationen charakterisiert. Dabei konnte erstmals die Existenz einer stabilen Synchronisationsmannigfaltigkeit für derartig komplexe Dynamik nachgewiesen werden. Es stellte sich heraus, dass die Synchronisation mit hervorragenden *Chaos-Pass-Filter* (CPF)-Eigenschaften verbunden ist, die sich darin äußern, dass das Empfängersystem zwischen dem chaotischen Trägersignal und einer Nachricht unterscheidet, die in Form einer Störung des Trägersignals verschlüsselt werden kann. Die exzellenten Synchronisations- und CPF-Eigenschaften beider Konfigurationen wurden in Hinblick auf Sicherheits- und Zuverlässigkeitsaspekte möglicher Chaos-Kommunikationssysteme bewertet. Unter diesen Aspekten wurden zwei unterschiedliche Chaos-Kommunikationsschemata untersucht. Zum einen wurde das neu entwickelte *ON/OFF-Phase-Shift-Keying* Schema vorgestellt, das ein hohes Maß an Sicherheit bietet. Zum anderen wurde ein voll funktionstüchtiges Chaos-Kommunikationssystem realisiert, das auf einem optimierten *Chaos Shift Keying* Schema basiert. Für dieses System konnte die erfolgreiche Übertragung einer digitalen Nachricht demonstriert werden. Dieses realisierte System vereint ein hohes Maß an Sicherheit mit zuverlässiger Funktionalität bei hohen Übertragungsraten bis hin zu einigen Gbit/s. Abschließend wurde eine vergleichende Übersicht der Eigenschaften unterschiedlicher Konfigurationen bereitgestellt, die als Basis für eine Weiterentwicklung dieser fruchtbaren Anwendung von Chaos-Synchronisation dienen kann.

Zusammenfassend kann festgestellt werden, dass die präsentierten Ergebnisse die enge Verknüpfung zwischen NLD und HLL-Physik herausgestellt haben. Einerseits wurde nachgewiesen, dass HLL-Systeme als vielseitige und gut kontrollierbare Modellsysteme zur experimentellen Untersuchung grundlegender Phänomene der NLD dienen können. Andererseits konnte die NLD erfolgreich zur Beeinflussung der dynamischen

Eigenschaften von HL-Systemen genutzt werden. Dies zeigte im Gegenzug, dass sich durch angewandte NLD neue Anwendungsbereiche von HLL erschließen lassen. Die präsentierten Ergebnisse heben die nützliche wechselseitige Beziehung zwischen NLD und HL-Physik hervor, die als fruchtbare Basis für zukünftige Fortschritte in beiden Bereichen der Wissenschaft dienen kann.

Appendix

A.1 Specifications of the used Lasers

Table A.1 summarizes relevant specifications and properties of the SLs which are used in the experiments presented in this thesis.

The data is presented in tabular form to allow for simple comparison of the characteristics. On the one hand, the table comprises characteristics which are (dominantly) related to resonator specific properties: resonator type; geometric device length, L_{SL}; side mode suppression, P_{side}/P_{center} for intermediate output power. On the other hand, the table outlines specifications which are (prevailingly) related to material proper-

	Semiconductor Laser			
	HLP1400	**FBH**	**Optospeed**	**IPAG**
Resonator	Fabry-Perot	Fabry-Perot	DFB	DFB
L_{SL}	300 μm	1.6 mm	300 μm	300 μm
P_{side}/P_{center}	(-15 dB)	(-20 dB)	-47 dB	-43 dB
Polarization	TE	TM	TE	TE
λ_{center} @ 22° C	839 nm	786 nm	1541 nm	1541 nm
n_{eff}	3.6	3.7	3.5	3.5
$\Delta\nu_{LM}$	138.9 GHz	25.3 GHz	144.5 GHz	144.5 GHz
$I_{th,sol}$ @ 22° C	64.1 mA	45.2 mA	5.7 mA	7.4 mA
$\Delta P/\Delta I_{DC}$	0.32 W/A	0.96 W/A	0.22 W/A	0.25 W/A
P_{max}	14.7 mW	110 mW	21 mW	23 mW
$\nu_{RO,p=1.2}$	2.4 GHz	0.9 GHz	1.9 GHz	2.1 GHz
$\nu_{RO,max}$	4.3 GHz	3.2 GHz	17.2 GHz	17.0 GHz
α	2.7	2.0	2.5	2.5

Table A.1: Characteristics of the different semiconductor lasers utilized in the experiments.

ties: polarization; center wavelength for solitary operation under normal conditions, λ_{center}@22° C; effective refractive index of the gain material, n_{eff}; longitudinal mode spacing, $\Delta\nu_{LM}$; solitary threshold current, $I_{th,sol}$@22° C; slope efficiency, $\Delta P/\Delta I_{DC}$; and maximum output power, P_{max}. Finally, the table highlights dynamics characteristics which are fundamentally influenced by both, resonator and gain material of the SLs: relaxation oscillation frequency for moderate and maximum pump parameter, $\nu_{RO,p=1.2}$, $\nu_{RO,max}$; linewidth enhancement factor at center emission wavelength, α-parameter.

A.2   Perfect and Generalized Synchronization

Perfect Synchronization of two systems considerably differs from *General Synchronization*. Characteristic for perfect synchronization, or identical synchronization, is that it requires two identical oscillators 1 and 2 whose state variables, $\mathbf{x}_1$ and $\mathbf{x}_2$, can fulfill the following synchronization condition [162]:

$$\lim_{t\to\infty} ||\mathbf{x}_1(t) - \mathbf{x}_2(t + \Delta t)|| = 0$$

Here, Δt denotes a constant in time. In principle, for the closed loop configuration this synchronization condition can be realized if two identical lasers are utilized that are weakly coupled. In this case, the intensities which are coupled into each of the lasers are required to agree to achieve corresponding dynamical properties for both lasers. Additionally, it is supposed that the remaining laser parameters are identical. For these conditions, the intensity dynamics of both lasers can perfectly synchronize:

$$I_2(t + \Delta t) = I_1(t) \tag{A.1}$$

This condition has been intensively studied by numerical modeling [195, 196, 203]. In contrast to detailed results for perfect synchronization obtained from numerical modeling, successful experimental investigation of perfect synchronization is rare. This is due to the fact that experiments on perfect synchronization are very challenging because the allowed tolerances for perfect synchronization are very small ($< 1\%$). However, recently Liu et. al. gave first evidence for perfect synchronization of SL systems [200]. Experimentally, perfect synchronization is verified by analysis of the time delay of the synchronized response system 2 with respect to the driving system 1. In this analysis the coupling time τ_c and the feedback time τ_{EC} need to be considered. Perfect synchronization is achieved when the intensity dynamics of both laser systems suffices the following relation:

$$I_2(t + \tau_c - \tau_{EC}) = I_1(t) \tag{A.2}$$

In the closed and in the open loop configuration, measurements of the time delay between the drive and the response systems for moderate coupling conditions usually reveal time delays corresponding to $\Delta t = \tau_c$. This solution is typical for general synchronization for which the drive system 1 and the response system 2 fulfill a generalized dependence, which might be either described by a functional relation H or by an attractor:

$$\lim_{t \to \infty} H(\mathbf{x}_2(t) - \mathbf{x}_1(t)) = 0 \tag{A.3}$$

Experimentally, such a generalized relation can be easily verified by plotting the intensity $I_2(t + \tau_c)$ of the response system 2 against the time series $I_1(t)$ of the drive system 1. If there is a relation between both laser systems, then the resulting figure often reveals a structure corresponding to a functional relation h between the dynamics of both laser systems:

$$h(I_1(t)) \times I_2(t + t_c) = I_1(t) \tag{A.4}$$

Most of the experiments presented in this thesis reveal almost identical intensity dynamics for both of the two laser systems. Hence, in the experiments $h(I_1(t))$ is almost constant.

Bibliography

[1] H. Poincaré, *Les Méthodes Nouvelles de la Méchanique Celeste*, vol. 1 and 2 (Gauthiers-Villars, Paris, 1892-1899).

[2] E. N. Lorenz, "Deterministic nonperiodic flow," J. Atmos. Sci. **20**(2), 130–141 (1963).

[3] M. Henon and C. Heiles, "Applicability of 3 integral of motion - some numerical experiments," Astron. J. **69**(1), 73–79 (1964).

[4] D. Ruelle and F. Takens, "Nature of turbulence," Commun. Math. Phys. **20**(3), 167–192 (1971).

[5] M. J. Feigenbaum, "Onset spectrum of turbulence," Phys. Lett. A **74**(6), 375–378 (1979).

[6] Y. Pomeau and P. Manneville, "Intermittent transition to turbulence in dissipative dynamical-systems," Commun. Math. Phys. **74**, 189–197 (1980).

[7] C. Grebogi, E. Ott, and J. A. Yorke, "Chaotic attractors in crisis," Phys. Rev. Lett. **48**(22), 1507–1510 (1982).

[8] K. T. Alligood, T. Sauer, and J. A. Yorke, *Chaos: An Introduction to Dynamical Systems* (Springer, New York, 2000).

[9] S. H. Strogatz, *Nonlinear Dynamics and Chaos: With Applications to Physics, Biology, Chemistry and Engineering* (Perseus Publishing, Cambridge, Massachusets, 2000).

[10] H. Haken, *Information and Self-Organization. A Macroscopic Approach to Complex Systems*, 2nd ed. (Springer, Berlin, 2000).

[11] H. Haken, "Analogy between higher instabilities in fluids and lasers," Phys. Lett. A **53**(1), 77–78 (1975).

[12] L. W. Casperson, "Spontaneous coherent pulsations in laser oscillators," IEEE J. Quantum Electron. **QE-14**(10), 756–761 (1978).

[13] F. T. Arecchi, R. Meucci, G. P. Puccioni, and J. R. Tredicce, "Experimental evidence of subharmanic bifurcation, multistability, and turbulence in a Q-switched gas laser," Phys. Rev. Lett. **49**(17), 1217–1220 (1982).

[14] T. Midavaine, D. Dangoisse, and P. Glorieux, "Observation of chaos in a frequency-modulated CO_2 laser," Phys. Rev. Lett. **55**(19), 1989–1992 (1985).

[15] C. O. Weiss and R. Vilaseca, *Dynamics of Lasers* (Wiley - VCH, Weinheim, 1991).

[16] D. V. Ivanov, Y. I. Khanin, I. I. Matorin, and A. S. Pikovsky, "Chaos in a solid-state laser with periodically modulated losses," Phys. Lett. **89**(5), 229–230 (1982).

[17] E. Brun, B. Derighetti, D. Meier, R. Holzner, and M. Ravani, "Observation of order and chaos in a nuclear-spin flip laser," Journ. Opt. Soc. Am. B **2**(1), 156–167 (1985).

[18] H. G. Winful, Y. C. Chen, and J. M. Liu, "Frequency locking, quasi-periodicity, and chaos in modulated self-pulsing semiconductor-lasers," Appl. Phys. Lett. **48**(10), 616–618 (1986).

[19] C. Risch and C. Voumard, "Self-pulsation in output intensity and spectrum of GaAs-AlGaAs cw diode lasers coupled to a frequency-selective external optical cavity," J. Appl. Phys. **48**(5), 2083–2085 (1977).

[20] R. N. Hall, G. E. Fenner, J. D. Kingsley, T. J. Soltys, and R. O. Carlson, "Coherent light emission from GaAs junctions," Phys. Rev. Lett. **9**(9), 366–369 (1962).

[21] M. I. Nathan, W. P. Dumke, G. Burns, F. H. Dill Jr., and G. Lasher, "Stimulated emission of radiation from GaAs p-n junctions," Appl. Phys. Lett **1**(3), 62–64 (1962).

[22] T. M. Quist, R. H. Rediker, R. J. Keyes, W. E. Krag, B. Lax, A. L. McWhorter, and H. J. Zeigler, "Semiconductor maser of GaAs," Appl. Phys. Lett. **1**(4), 91–92 (1962).

[23] J. Sacher, W. Elsäßer, and E. O. Göbel, "Intermittency in the coherence collapse of a semiconductor-laser with external feedback," Phys. Rev. Lett. **63**(20), 2224–2227 (1989).

[24] T. Erneux, F. Rogister, A. Gavrielides, and V. Kovanis, "Bifurcation to mixed external cavity mode solutions for semiconductor lasers subject to optical feedback," Opt. Commun. **183**(5–6), 467–477 (2000).

[25] J. Mørk, B. Tromborg, and J. Mark, "Chaos in semiconductor-lasers with optical feedback - theory and experiment," IEEE J. Quantum Electron. **QE-28**(1), 93–108 (1992).

[26] J. Ye, H. Li, and J. G. McInerney, "Period-doubling route to chaos in a semiconductor laser with weak optical feedback," Phys. Rev. A **47**(3), 2249–2252 (1993).

[27] G. C. Dente, P. S. Durkin, K. A. Wilson, and C. E. Moeller, "Chaos in the coherence collapse of semiconductor-lasers," IEEE J. Quantum Electron. **QE-24**(12), 2441–2447 (1988).

[28] F. Schweitzer, *Self-Organization of Complex Structes: From Individual to Collective Dynamics* (Overseas Publisher Association, Amsterdam, 1997).

[29] M. I. Rabinovich, A. B. Ezersky, and P. D. Weidman, *The Dynamics of Patterns* (World Scientific Publishing GroupCo. Pte. Ltd., Singapore, 2000).

[30] E. Schöll, *Nonlinear Spatio-Temporal Dynamics and Chaos in Semiconductors* (Cambridge University Press, Cambridge, 2001).

[31] A. Pikovsky, M. Rosenblum, and J. Kurths, *Synchronization: A Universal Concept in Nonlinear Sciences* (Cambridge University Press, Cambridge, 2003).

[32] S. H. Strogatz, *Synch: How Order Emerges from Chaos in the Universe, Nature, and Daily Life* (Hyperion, New York, 2004).

[33] E. Ott, T. Sauer, and J. A. York, *Coping with Chaos* (John Wiley & Sons, Inc., New York, 1994).

[34] L. A. Coldren and S. W. Corzine, *Diode Lasers and Photonic Integrated Circuits* (John Wiley & Sons, Inc., New York, 1995).

[35] T. Mukai and K. Otsuka, "New route to optical chaos: Successive subharmonic oscillation cascade in a semiconductor laser coupled to an external cavity," Phys. Rev. Lett. **55**(17), 1711–1714 (1985).

[36] K. Ikeda and K. Matsumoto, "High-dimensional chaotic behavior in systems with time-delayed feedback," Physica D **29**(1-2), 223–235.

[37] I. Fischer, O. Hess, W. Elsäßer, and E. Göbel, "High-dimensional chaotic dynamics of an external-cavity semiconductor laser," Phys. Rev. Lett. **73**(16), 2188–2191 (1994).

[38] A. Yariv, *Quantum Electronics*, 3rd ed. (John Wiley & Sons, Inc., New York, 1989).

[39] J. D. Jackson, *Classical Electrodynamics*, 3rd ed. (John Wiley & Sons, Inc., New York, 1998).

[40] G. H. M. van Tartwijk, *Semiconductor Laser Dynamics with Optical Injection and Feedback* (Vrije Universiteit Amsterdam, 1994).

[41] H. Haken, *Light, Vol. II, Laser Light Dynamics* (North-Holland, Amsterdam, 1985).

[42] C. O. Weiss and J. Brock, "Evidence for lorenz-type chaos in a laser," Phys. Rev. Lett. **57**(22), 2804–2806 (1986).

[43] J. Sacher, D. Baums, P. Panknin, W. Elsäßer, and E. Göbel, "Intensity instabilities of semiconductor lasers under current modulation, external light injection, and delayed feedback," Phys. Rev. A **45**(3), 1893–1905 (1992).

[44] H. F. Liu and W. F. Ngai, "Nonlinear dynamics of a directly modulated 1.55 μm InGaAsP distributed feedback semiconductor laser," IEEE J. Quantum Electron. **QE-29**(6), 1668–1675 (1993).

[45] T. B. Simpson, J. M. Liu, A. Gavrielides, V. Kovanis, and P. M. Alsing, "Period-doubling route to chaos in a semiconductor laser subject to optical injection," Appl. Phys. Lett. **64**(26), 3539–3541 (1994).

[46] N. A. Loiko and A. M. Samson, "Possible regimes of generation of a semiconductor laser with a delayed optoelectric feedback," Opt. Commun. **93**(1-2), 66–72 (1992).

[47] J. Mørk, J. Mark, and B. Tromborg, "Route to chaos and competition between relaxation oscillations for a semiconductor laser with optical feedback," Phys. Rev. Lett. **65**(16), 1999–2002 (1990).

[48] G. P. Agrawal and N. K. Dutta, *Long-Wavelength Semiconductor Lasers* (Van Nostrand Reinhold Company, New York, 1993).

[49] T. Heil, *Ph.D. Thesis: Delay Dynamics in Semiconductor Lasers: Feedback and Coupling Induced Instabilities, Stabilization, and Synchronization* (Shaker Verlag, Aachen, 2001).

[50] G. Lasher and F. Stern, "Spontaneous and Stimulated Recombination Radiation in Semiconductors," Phys. Rev. **133**(2A), A533–A563 (1964).

[51] C. H. Henry, R. A. Logan, and K. A. Bertness, "Spectral dependence of the change in refractive index due to carrier injection in GaAs lasers," J. Appl. Phys. **52**(7), 4457–4461 (1981).

[52] H. A. Kramers, "La Diffusion de la lumière par les atomes." Atti. Congr. Int. Fis. Como **2**, 545–557 (1927).

[53] R. L. de Kronig, "On the theory of dispersion of X-rays," J. Opt. Soc. Am. **12**, 547–557 (1926).

[54] H. Haug and H. Haken, "Theory of noise in semiconductor laser emission," Z. Physik **204**(3), 262–275 (1967).

[55] C. H. Henry, "Theory of the linewidth of semiconductor lasers," IEEE J. Quantum Electron. **QE-18**(2), 259–264 (1982).

[56] Y. Arakawa and A. Sakaki, "Multidimensional quantum well laser and temperature dependence of its threshold current," Appl. Phys Lett. **40**(11), 939–941 (1982).

[57] J. Faist, F. Capasso, D. L. Sivco, C. Sirtori, A. L. Hutchinson, and A. Y. Cho, "Quantum cascade laser," Science **264**(5158), 553–556 (1994).

[58] K. Vahala and A. Yariv, "Semiclassical theory of noise in semiconductor lasers – Part I," IEEE J. Quantum Electron. **QE-19**(6), 1096–1101 (1983).

[59] M. Osinski and J. Buus, "Linewidth broadening factor in semiconductor lasers – an overview," IEEE J. Quantum Electron. **QE-23**(1), 9–28 (1987).

[60] B. W. Hakki and T. L. Paoli, "Gain spectra in GaAs double–heterostructure injection lasers," J. Appl. Phys. **46**(3), 1299–1306 (1975).

[61] I. D. Henning and J. V. Collins, "Measurements of the semiconductor laser linewidth broadening factor," Electron. Lett. **19**(22), 927–929 (1983).

[62] M. P. van Exter and J. P. Woerdman, "Determination of alpha factor of fabry-perot-type semiconductor laser by injection locking," Electron. Lett. **28**(17), 1607–1608 (1992).

[63] Y. Yu, G. Giuliani, and S. Donati, "Measurement of the linewidth enhancement factor of semiconductor lasers based on the optical feedback self-mixing effect," IEEE Photon. Technol. Lett. **PTL-16**(4), 990–992 (2004).

[64] L. Olofsson and T. G. Brown, "The influence of resonator structure on the linewidth enhancement factor of semiconductor lasers," IEEE J. Quantum Electron. **QE-28**(6), 1450–1458 (1992).

[65] K. Vahala, L. C. Chiu, S. Margalit, and A. Yariv, "On the linewidth enhancement factor α in semiconductor injection lasers," Appl. Phys. Lett **42**(8), 631–633 (1983).

[66] I. Fischer, O. Hess, W. Elsäßer, and E. Göbel, "Complex spatio-temporal dynamics in the near-field of a broad-area semiconductor laser," Europhys. Lett. **35**(8), 579–584 (1996).

[67] A. Barchanski, T. Gensty, C. Degen, I. Fischer, and W. Elsäßer, "Picosecond emission dynamics of vertical-cavity surface- emitting lasers: Spatial, spectral, and polarization-resolved characterization," IEEE J. Quantum Electron. **QE-39**(7), 850–858 (2003).

[68] T. Gensty, K. Becker, I. Fischer, W. Elsäßer, C. Degen, P. Debernardi, and G. P. Bava, "Wave chaos in real-world vertical-cavity surface-emitting lasers," Phys. Rev. Lett. **94**(23), 233901 (2005).

[69] J. Wieland, C. R. Mirasso, and D. Lenstra, "Prevention of coherence collapse in diode lasers by dynamic targeting," Opt. Lett. **22**(7), 469–471 (1997).

[70] T. Heil, I. Fischer, and W. Elsäßer, "Stabilization of feedback-induced instabilities in semiconductor lasers," J. Opt. B: Quantum Semiclass. Opt. **2**(3), 413–420 (2000).

[71] F. Y. Lin and H. M. Liu, "Chaotic lidar," IEEE J. Sel. Topics Quantum Electron. **STQE-10**(5), 991–997 (2004).

[72] D. Lenstra, B. H. Verbeek, and A. J. den Boef, "Coherence collapse in single-mode semiconductor lasers due to optical feedback," IEEE J. Quantum Electron. **QE-21**(6), 674–679 (1985).

[73] J. Mørk and B. Tromborg, "The mechanism of mode selection for an external cavity laser," IEEE Photon. Technol. Lett. **PTL-2**(1), 21–23 (1990).

[74] K. Petermann, "External optical feedback phenomena in semiconductor-lasers," IEEE J. Select. Topics Quantum Electron. **STQE-1**(2), 480–489 (1995).

[75] T. L. Paoli and J. E. Ripper, "Frequency stabilization and narrowing of optical pulses from cw GaAs injection lasers," IEEE J. Quantum Electron. **QE-6**(6), 335–339 (1970).

[76] G. P. Agrawal, "Line narrowing in a single-mode injection-laser due to external optical feedback," IEEE J. Quantum Electron. **QE-20**(5), 468–471 (1984).

[77] H. Olesen, J. H. Osmundsen, and B. Tromborg, "Nonlinear dynamics and spectral behavior for an external cavity laser," IEEE J. Quantum Electron. **QE-22**(6), 762–773 (1986).

[78] C. R. Mirasso, P. Colet, and P. Garcia-Fernandez, "Synchronization of chaotic semiconductor lasers: Application to encoded communications," IEEE Photon. Technol. Lett. **PTL-8**(2), 299–301 (1996).

[79] V. Annovazzi-Lodi, S. Donati, and A. Scire, "Synchronization of chaotic injected-laser systems and its application to optical cryptography," IEEE J. Quantum Electron. **QE-32**(6), 953–959 (1996).

[80] I. Fischer, Y. Liu, and P. Davis, "Synchronization of chaotic semiconductor laser dynamics on subnanosecond time scales and its potential for chaos communication," Phys. Rev. A **62**(1), 011801 (2000).

[81] M. Peil, I. Fischer, and W. Elsäßer, "A short external cavity semiconductor laser cryptosystem," C. R. Physique **5**(6), 633–642 (2004).

[82] A. Argyris, D. Syvridis, L. Larger, V. Annovazzi-Lodi, P. Colet, I. Fischer, J. Garcia-Ojalvo, C. R. Mirasso, L. Pesquera, and K. A. Shore, "Chaos-based communications at high bit rates using commercial fibre-optic links," Nature **438**(17), 343–346 (2005).

[83] P. A. Ruprecht and J. R. Brandenberger, "Enhancing diode-laser tuning with a short external cavity," Opt. Commun. **93**(1–2), 82–86 (1992).

[84] R. W. Tkach and A. R. Chraplyvy, "Regimes of feedback effects in 1.5 μm distributed feedback lasers," J. Lightwave Technol. **LT-4**(11), 1655–1661 (1986).

[85] Y. Ikuma and J. Ohtsubo, "Dynamics in a compound cavity semiconductor laser induced by small external-cavity-length change," IEEE J. Quantum Electron. **QE-34**(7), 1240–1246 (1998).

[86] T. Heil, I. Fischer, W. Elsäßer, and A. Gavrielides, "Dynamics of semiconductor lasers subject to delayed optical feedback: The short cavity regime," Phys. Rev. Lett. **87**(24), 243901 (2001).

[87] T. Heil, I. Fischer, W. Elsäßer, B. Krauskopf, K. Green, and A. Gavrielides, "Delay dynamics of semiconductor lasers with short external cavities: Bifurcation scenarios and mechanisms," Phys. Rev. E **67**(6), 066214 (2003).

[88] R. Lang and K. Kobayashi, "External optical feedback effects on semiconductor injection-laser properties," IEEE J. Quantum Electron. **QE-16**(3), 347–355 (1980).

[89] T. Heil, I. Fischer, and W. Elsäßer, "Coexistence of low-frequency fluctuations and stable emission on a single high-gain mode in semiconductor lasers with external optical feedback," Phys. Rev. A **58**(4), R2672 (1998).

[90] I. Fischer, T. Heil, and W. Elsäßer, "Emission dynamics of semiconductor lasers subject to delayed optical feedback: An experimentalist's perspective," in *Fundamental Issues of Nonlinear Laser Dynamics*, B. Krauskopf and D. Lenstra, eds., Conference Proceedings Vol. 548, American Institute of Physics, Melville N.Y., pp. 218–237 (2000).

[91] J. Sacher, W. Elsäßer, and E. O. Göbel, "Nonlinear dynamics of semiconductor laser emission under variable feedback conditions," IEEE J. Quantum Electron. **QE-27**(3), 373–379 (1991).

[92] T. W. Carr, D. Pieroux, and P. Mandel, "Theory of a multimode semiconductor laser with optical feedback," Phys. Rev. A **63**(3), 033817 (2001).

[93] A. Uchida, Y. Liu, I. Fischer, P. Davis, and T. Aida, "Chaotic antiphase dynamics and synchronization in multimode semiconductor lasers," Phys. Rev. A **64**(2), 023801 (2001).

[94] A. Murakami, J. Ohtsubo, and Y. Liu, "Stability analysis of semiconductor laser with phase-conjugate feedback," IEEE J. Quantum Electron. **QE-33**(10), 1825–1831 (1997).

[95] D. H. DeTienne, G. R. Gray, G. P. Agrawal, and L. D, "Semiconductor laser dynamics for feedback from a finite-penetration-depth phase-conjugate mirror," IEEE J. Quantum Electron **QE-33**(5), 838–844 (1997).

[96] T. Heil, A. Uchida, P. Davis, and T. Aida, "TE-TM dynamics in a semiconductor laser subject to polarization-rotated optical feedback," Phys. Rev. A **68**(3), 033811 (2003).

[97] D. W. Sukow, K. L. Blackburn, A. R. Spain, K. J. Babcock, L. V. Bennett, and A. Gavrielides, "Experimental synchronization of chaos in diode lasers with polarization-rotated feedback and injection," Opt. Lett. **29**(20), 2393–2395 (2004).

[98] G. H. M. van Tartwijk, A. M. Levine, and D. Lenstra, "Sisyphus effect in semiconductor-lasers with optical feedback," IEEE J. Select. Topics Quantum Electron. **STQE-1**(2), 466–472 (1995).

[99] Y. Liu, P. Davis, and Y. Takiguchi, "Recovery process of low-frequency fluctuations in laser diodes with external optical feedback," Phys. Rev. E **60**(6), 6595–6601 (1999).

[100] G. Vaschenko, M. Giudici, J. J. Rocca, C. S. Menoni, J. R. Tredicce, and S. Balle, "Temporal dynamics of semiconductor lasers with optical feedback," Phys. Rev. Lett **81**(25), 5536–5539 (1998).

[101] D. W. Sukow, T. Heil, I. Fischer, A. Gavrielides, A. Hohl-AbiChedid, and W. Elsäßer, "Picosecond intensity statistics of semiconductor lasers operating in the low-frequency fluctuation regime," Phys. Rev. A **60**(1), 667–673 (1999).

[102] M. Peil, I. Fischer, and W. Elsäßer, "Semiconductor laser system: Dynamics beyond Lang-Kobayashi," in *Proceedings of ENOC-2005, Fifth EUROMECH Nonlinear Dynamics Conference*, D. H. van Campen, M. D. Lazurko, and W. P. J. M. van den Oever, eds., MS 18-269, pp. 2074–2082 (2005).

[103] C. Szwaj, S. Bielawski, D. Derozier, and T. Erneux, "Faraday instability in a multimode laser," Phys. Rev. Lett. **80**(18), 3968–3971 (1998).

[104] T. Sano, "Antimode dynamics and chaotic itinerancy in the coherence collapse of semiconductor lasers with optical feedback," Phys. Rev. A **50**(3), 2719–2726 (1994).

[105] B. Tromborg, J. Osmundsen, and H. Olesen, "Stability analysis for a semiconductor laser in an external cavity," IEEE J. Quantum Electron. **QE-20**(9), 1023–1032 (1984).

[106] A. M. Levine, G. H. M. Van Tartwijk, D. Lenstra, and T. Erneux, "Diode-lasers with optical feedback - stability of the maximum gain mode," Phys. Rev. A **52**(5), R3436 (1995).

[107] I. Fischer, G. H. M. vanTartwijk, A. M. Levine, W. Elsäßer, E. Göbel, and D. Lenstra, "Fast pulsing and chaotic itinerancy with a drift in the coherence collapse of semiconductor lasers," Phys. Rev. Lett. **76**(2), 220–223 (1996).

[108] A. Hohl, H. J. C. van der Linden, and R. Roy, "Determinism and stochasticity of power-dropout events in semiconductor lasers with optical," Opt. Lett. **20**(23), 2396–2398 (1995).

[109] J. M. Buldú, J. Garcia-Ojalvo, C. R. Mirasso, and M. C. Torrent, "Stochastic entrainment of optical power dropouts," Phys. Rev. E **66**(2), 021106 (2002).

[110] O. Hess and T. Kuhn, "Maxwell-Bloch equations for spatially inhomogeneous semiconductor lasers.1. Theoretical formulation," Phys. Rev. A **54**(4), 3347–3359 (1996).

[111] B. Tromborg, J. Mørk, and V. Velichansky, "On mode coupling and low-frequency fluctuations in external-cavity laser diodes," Quantum Semiclass. Opt. **9**(5), 831–851 (1997).

[112] M. Münkel, F. Kaiser, and O. Hess, "Stabilization of spatiotemporally chaotic semiconductor laser arrays by means of delayed optical feedback," Phys. Rev. E **56**(4), 3868–3875 (1997).

[113] J. K. White and J. V. Moloney, "Multichannel communication using an infinite dimensional spatiotemporal chaotic system," Phys. Rev. A **59**(3), 2422–2426 (1999).

[114] M. Möhrle, B. Sartorius, C. Bornholdt, S. Bauer, O. Brox, A. Sigmund, R. Steingrüber, M. Radziunas, and H.-J. Wünsche, "Detuned grating multisection-RW-DFB lasers for high-speed optical signal processing," IEEE J. Select. Topics Quantum Electron. **STQE-7**(2), 217–223 (2001).

[115] A. T. Ryan, G. P. Agrawal, G. R. Gray, and E. C. Gage, "Optical-feedback-induced chaos and its control in multimode semiconductor lasers," IEEE J. Quantum Electron. **QE-30**(3), 668–679 (1994).

[116] E. A. Viktorov and P. Mandel, "Transmission of encoded information based on antiphase dynamics," Opt. Lett. **22**(20), 1568–1570 (1997).

[117] J. Mørk, B. Tromborg, and P. L. Christiansen, "Bistability and low-frequency fluctuations in semiconductor-lasers with optical feedback - a theoretical-analysis," IEEE J. Quantum Electron. **QE-24**(2), 123–133 (1988).

[118] D. Lenstra, "Fundamental nonlinear dynamics of semiconductor lasers," Quantum Semiclass. Opt. **9**(5), U3–U5 (1997).

[119] T. Heil, I. Fischer, and W. Elsäßer, "Influence of amplitude-phase coupling on the dynamics of semiconductor lasers subject to optical feedback," Phys. Rev. A **60**(1), 634–641 (1999).

[120] E. A. Viktorov and P. Mandel, "Low frequency fluctuations in a multimode semiconductor laser with optical feedback," Phys. Rev. Lett. **85**(15), 3157–3160 (2000).

[121] F. Rogister, P. Megret, O. Deparis, and M. Blondel, "Coexistence of in-phase and out-of-phase dynamics in a multimode external-cavity laser diode operating in the low- frequency fluctuations regime," Phys. Rev. A **62**(6), 061803(R) (2000).

[122] D. J. Yu, I. Wallace, R. G. Harrison, and A. Gavrielides, "Low frequency fluctuations and locked states in a multi-mode semiconductor laser with external cavity," Opt. Commun. **195**(1–4), 249–258 (2001).

[123] F. Rogister, M. Sciamanna, O. Deparis, P. Megret, and M. Blondel, "Low-frequency fluctuation regime in a multimode semiconductor laser subject to a mode-selective optical feedback," Phys. Rev. A **65**(1), 015602 (2002).

[124] M. Yousefi, A. Barsella, D. Lenstra, G. Morthier, R. Baets, S. McMurtry, and J. P. Vilcot, "Rate equations model for semiconductor lasers with multilongitudinal mode competition and gain dynamics," IEEE J. Quantum Electron. **QE-39**(10), 1229–1237 (2003).

[125] I. V. Koryukin and P. Mandel, "Dynamics of semiconductor lasers with optical feedback: Comparison of multimode models in the low-frequency fluctuation regime," Phys. Rev. A **70**(5), 053819 (2004).

[126] C. L. Tang, H. Statz, and G. deMars, "Spectral output and spiking behavior of solid-state lasers," J. Appl. Phys. **34**(8), 2289–2295 (1963).

[127] P. Mandel, E. A. Viktorov, C. Masoller, and M. S. Torre, "Antiphase dynamics in a multimode Fabry-Perot semiconductor laser with external feedback," Physica A **327**(1-2), 129–134 (2003).

[128] M. Münkel, F. Kaiser, and O. Hess, "Spatio-temporal dynamics in semiconductor lasers with delayed optical feedback," Int. J. Bif. Chaos **8**(5), 951–963 (1998).

[129] A. A. Duarte and H. G. Solari, "Modeling the spatio-temporal dynamics of semiconductor lasers: the monochromatic solutions," Opt. Commun. **144**(1-3), 99–108 (1997).

[130] M. Yousefi, D. Lenstra, and G. Vemuri, "Carrier inversion noise has important influence on the dynamics'of a semiconductor laser," IEEE J. Sel. Topics Quantum Electron. **10**(5), 955–960 (2004).

[131] S. Yanchuk and M. Wolfrum, "Instabilities of stationary states in lasers with long-delay optical feedback," submitted (2004).

[132] M. P. van Exter, R. F. M. Hendriks, J. P. Woerdman, and C. J. van Poel, "Explanation of double-peaked intensity noise spectrum of an external-cavity semiconductor laser," Opt. Commun. **110**(1–2), 137–140 (1994).

[133] M. Sondermann and T. Ackemann, "Correlation properties and drift phenomena in the dynamics of vertical-cavity surface-emitting lasers with optical," Opt. Express **13**(7), 2707–2715 (2005).

[134] V. Ahlers, U. Parlitz, and W. Lauterborn, "Hyperchaotic dynamics and synchronization of external-cavity semiconductor lasers," Phys. Rev. E **58**(6), 7208–7213 (1998).

[135] M. Le Berre, E. Ressayre, A. Tallet, H. M. Gibbs, D. L. Kaplan, and M. H. Rose, "Conjecture on the dimensions of chaotic attractors of delayed-feedback dynamical systems," Phys. Rev. A **35**(9), 4020–4022 (1987).

[136] W. A. Hamel, M. P. van Exter, and J. Woerdman, "Coherence properties of a semiconductor-laser with feedback from a distant reflector - experiment and theory," IEEE J. Quantum Electron. **QE-28**(6), 1459–1469 (1992).

[137] M. R. Daza, A. Tarun, K. Fujita, and C. Saloma, "Temporal coherence behavior of a semiconductor laser under strong optical feedback," Opt. Commun. **161**(1), 123–131 (1999).

[138] C. Serrat, S. Prins, and R. Vilaseca, "Dynamics and coherence of a multimode semiconductor laser with optical feedback in an intermediate-length external-cavity regime," Phys. Rev. A **68**(5), 053804 (2003).

[139] W. Elsäßer and E. O. Göbel, "Multimode Effects in the spectral linewidth of semiconductor lasers," IEEE J. Quantum Electron. **QE-21**(6), 687–691 (1985).

[140] R. Vicente, J. Dauden, P. Colet, and R. Toral, "Analysis and characterization of the hyperchaos generated by a semiconductor laser subject to a delayed feedback loop," IEEE J. Quantum Electron. **QE–41**(4), 541–548 (2005).

[141] J. D. Farmer, "Chaotic attractors of an infinite-dimensional dynamical system," Physica D **4**(3), 366–393 (1982).

[142] H. Kantz and T. Schreiber, *Nonlinear Timer Series Analysis* (Cambridge University Press, Cambridge, U.K., 2000).

[143] A. Tabaka, K. Panajotov, I. Veretennicoff, and M. Sciamanna, "Bifurcation study of regular pulse packages in laser diodes subject to optical feedback," Phys. Rev. E **70**(3), 036211 (2004).

[144] M. Peil, T. Heil, I. Fischer, and W. Elsäßer, "Synchronization of chaotic semiconductor laser systems: A vectorial coupling-dependent scenario," Phys. Rev. Lett. **88**(17), 174101 (2002).

[145] J. B. Pesin, "Characteristic Lyapunov exponents and smooth ergodic theory," Russ. Math. Surveys **32**(4), 55–114 (1977).

[146] M. E. Brezinski and J. G. Fujimoto, "Optical coherence tomography: High-resolution imaging in nontransparent tissue," IEEE J. Select. Topics Quantum Electron. **STQE-5**(4), 1185–1192 (1999).

[147] P.-A. Champert, V. Couderc, P. Leproux, S. Février, V. Tombelaine, L. Labonté, P. Roy, and C. Froehly, "White-light supercontinuum generation in normally dispersive optical fiber using original multi-wavelength pumping system," Opt. Express **12**(19), 4366–4371 (2004).

[148] A. K. Abeeluck and C. Headley, "Supercontinuum growth in a highly nonlinear fiber with a low-coherence semiconductor laser diode," Appl. Phys. Lett. **85**(21), 4863–4865 (2004).

[149] L. Vaissié, O. V. Smolski, and E. G. Johnson, "Crossed-beam superluminescent diode," Opt. Lett. **30**(13), 1608–1610 (2005).

[150] C. H. Henry and R. F. Karzarinov, "Stabilization of single frequency operation of coupled-cavity lasers," IEEE J. Quantum Electron. **QE-20**(7), 733–744 (1984).

[151] D. Marcuse and T.-P. Lee, "Rate equation model of a coupled-cavity laser," IEEE J. Quantum Electron. **QE-20**(2), 166–176 (1984).

[152] L. Furfaro, F. Pedaci, M. Giudici, X. Hachair, J. Tredicce, and S. Balle, "Mode-switching in semiconductor lasers," IEEE J. Quantum Electron. **QE-40**(10), 1365–1376 (2004).

[153] A. M. Yacomotti, L. Furfaro, X. Hachair, F. Pedaci, M. Giudici, J. Tredicce, J. Javaloyes, S. Balle, E. A. Viktorov, and P. Mandel, "Dynamics of multimode semiconductor lasers," Phys. Rev. A **69**(5), 053816 (2004).

[154] W. Elsäßer, "Inter-intramode correlation spectroscopy: A new experimental technique to study mode interaction in semiconductor lasers," Appl. Phys. Lett. **48**(20), 1323–1325 (1986).

[155] P. Mandel, B. A. Nguyen, and K. Otsuka, "Universal dynamical properties of three-mode Fabry-Perot lasers," Quantum Semiclass. Opt. **9**(3), 365–380 (1997).

[156] J. M. Buldú, F. Rogister, J. Trull, C. Serrat, M. C. Torrent, J. Garcia-Ojalvo, and C. R. Mirasso, "Asymmetric and delayed activation of side modes in multimode semiconductor lasers with optical feedback," J. Opt. B: Quantum Semiclass. Opt. **4**(6), 415–420 (2002).

[157] C. Masoller, M. S. Torre, and P. Mandel, "Antiphase dynamics in multimode semiconductor lasers with optical feedback," Phys. Rev. A **71**(1), 013818 (2005).

[158] K. Otsuka and Y. Aizawa, "Gain circulation in multimode lasers," Phys. Rev. Lett. **72**(17), 2701–2704 (1994).

[159] V. S. Afraimovich, N. N. Verichev, and M. I. Rabinovich, "Stochastically synchronized oscillations in dissipative systems," Izv. Vyssh. Uchebn. Zaved. Radiofiz. **26**(9), 1050–1060 (1986).

[160] L. M. Pecora and T. L. Carroll, "Synchronization in chaotic systems," Phys. Rev. Lett. **64**(8), 821–824 (1990).

[161] H. G. Winful and L. Rahman, "Synchronized chaos and spatiotemporal chaos in arrays of coupled lasers," Phys. Rev. Lett. **65**(13), 1575–1578 (1990).

[162] M. G. Rosenblum, A. S. Pikovsky, and J. Kurths, "From phase to lag synchronization in coupled chaotic oscillators," Phys. Rev. Lett. **78**(22), 4193–4196 (1997).

[163] T. Sugawara, M. Tachikawa, T. Tsukamoto, and T. Shimizu, "Observation of synchronization in laser chaos," Phys. Rev. Lett. **72**(22), 3502–3505 (1994).

[164] R. Roy and K. S. Thornburg, "Experimental synchronization of chaotic lasers," Phys. Rev. Lett. **72**(13), 2009–2012 (1994).

[165] K. Coffman, W. D. McCormick, and H. L. Swinney, "Multiplicity in a chemical-reaction with one-dimensional dynamics," Phys. Rev. Lett **56**(10), 999–1002 (1986).

[166] S. H. Strogatz and I. Stewart, "Coupled oscillators and biological synchronization," Sci. Am. **269**(6), 102–109 (1993).

[167] C. Schäfer, M. G. Rosenblum, H. H. Abel, and J. Kurths, "Synchronization in the human cardiorespiratory system," Phys. Rev. E **60**(1), 857–870 (1999).

[168] L. Glass, "Synchronization and rhythmic processes in physiology," Nature **410**(6825), 277–284 (2001).

[169] F. C. Hoppensteadt and E. M. Izhikevich, "Synchronization of laser oscillators, associative memory, and optical neurocomputing," Phys. Rev. E **62**(3), 4010–4013 (2000).

[170] G. D. VanWiggeren and R. Roy, "Optical communication with chaotic waveforms," Phys. Rev. Lett **81**(16), 3547–3550 (1998).

[171] K. M. Cuomo and A. V. Oppenheim, "Circuit implementation of synchronized chaos with applications to communications," Phys. Rev. Lett. **71**(1), 65–68 (1993).

[172] C. R. Mirasso, J. Mulet, and C. Masoller, "Chaos shift-keying encryption in chaotic external-cavity semiconductor lasers using a single-receiver scheme," IEEE Photon. Technol. Lett. **PTL-14**(4), 456–458 (2002).

[173] G. D. van Wiggeren and R. Roy, "Communication with dynamically fluctuating states of light polarization," Phys. Rev. Lett **88**(9), 097903 (2002).

[174] L. Larger, J.-P. Goedgebuer, and F. Delorme, "Optical encryption system using hyperchaos generated by an optoelectronic wavelength oscillator," Phys. Rev. E **57**(6), 6618–6624 (1998).

[175] T. Heil, J. Mulet, I. Fischer, C. R. Mirasso, M. Peil, P. Colet, and W. Elsäßer, "ON/OFF phase shift keying for chaos-encrypted communication using external-cavity semiconductor lasers," IEEE J. Quantum Electron. **Q-38**(9), 1162–1170 (2002).

[176] A. Uchida, S. Yoshimori, M. Shinozuka, T. Ogawa, and F. Kannari, "Chaotic on-off keying for secure communications," Opt. Lett. **26**(12), 866–868 (2001).

[177] G. D. VanWiggeren and R. Roy, "Chaotic communication using time-delayed optical systems," Int. J. Bif. Chaos **9**(11), 2129–2156 (1999).

[178] J. B. Geddes, K. M. Short, and K. Black, "Extraction of signals from chaotic laser data," Phys. Rev. Lett. **83**(25), 5389–5392 (1999).

[179] K. M. Short and A. T. Parker, "Unmasking a hyperchaotic communication scheme," Phys. Rev. E **58**(1), 1159–1162 (1998).

[180] V. S. Udaltsov, J.-P. Goedgebauer, L. Larger, and W. T. Rhodes, "Communicating with optical hyperchaos: Information encryption and decryption in delayed nonlinear feedback systems," Phys. Rev. Lett. **86**(9), 1892–1895 (2001).

[181] V. S. Udaltsov, J.-P. Goedgebuer, L. Larger, J. B. Cuenot, P. Levy, and W. T. Rhodes, "Cracking chaos-based encryption systems ruled by nonlinear time delay differential equations," Phys. Lett. A **308**(1), 54–60 (2003).

[182] Y. Takiguchi, H. Fujino, and J. Ohtsubo, "Experimental synchronization of chaotic oscillations in externally injected semiconductor lasers in a low-frequency fluctuation regime," Opt. Lett. **24**(22), 1570–1572 (1999).

[183] H. Fujino and J. Ohtsubo, "Experimental synchronization of chaotic oscillations in external-cavity semiconductor lasers," Opt. Lett. **25**(9), 625–627 (2000).

[184] L. Wallace, D. J. Yu, W. P. Lu, and R. G. Harrison, "Synchronization of power dropouts in coupled semiconductor lasers with external feedback," Phys. Rev A **63**(1), 013809 (2000).

[185] J.-P. Goedgebuer, L. Larger, and H. Porte, "Optical cryptosystem based on synchronization of hyperchaos generated by a delayed feedback tunable laser diode," Phys. Rev. Lett. **80**(10), 2249–2252 (1998).

[186] D. Kanakidis, A. Argyris, and D. Syvridis, "Performance characterization of high-bit-rate optical chaotic communication systems in a back-to-back configuration," IEEE J. Lightwave Technol. **LT-21**(3), 750–758 (2003).

[187] S. Sivaprakasam and K. A. Shore, "Message encoding and decoding using chaotic external-cavity diode lasers," IEEE J. Quantum Electron. **QE-36**(1), 35–39 (2000).

[188] J. M. Liu, H. F. Chen, and S. Tang, "Synchronized chaotic optical communications at high bit rates," IEEE J. Quantum Electron. **QE-38**(9), 1184–1196 (2002).

[189] J. Ohtsubo, "Chaos synchronization and chaotic signal masking in semiconductor lasers with optical feedback," IEEE J. Quantum Electron. **QE-38**(9), 1141–1154 (2002).

[190] M. W. Lee, L. Larger, and J.-P. Goedgebuer, "Transmission system using chaotic delays between lightwaves," IEEE J. Quantum Electron. **QE-39**(7), 931–935 (2003).

[191] H. Fujisaka and T. Yamada, "Stability theory of synchronized motion in coupled-oscillator systems," Prog. Theor. Phys. **69**(1), 32–47 (1983).

[192] M. G. Rosenblum, A. S. Pikovsky, and J. Kurths, "Phase synchronization of chaotic oscillators," Phys. Rev. Lett. **76**(11), 1804–1807 (1996).

[193] H. D. I. Abarbanel, N. F. Rulkov, and M. M. Sushchik, "Generalized synchronization of chaos: The auxiliary system approach," Phys. Rev. E **53**(5), 4528–4535 (1996).

[194] F. Dachselt and W. Schwarz, "Chaos and cryptography," IEEE Trans Circuits Syst. I **CAS1-48**(12), 1498–1509 (2001).

[195] A. Locquet, F. Rogister, M. Sciamanna, P. Mégret, and M. Blondel, "Two types of synchronization in unidirectionally coupled chaotic external-cavity semiconductor lasers," Phys. Rev. E **64**(4), 045203 (2001).

[196] C. Masoller, "Anticipation in the synchronization of chaotic semiconductor lasers with optical feedback," Phys. Rev. Lett. **86**(13), 2782–2785 (2001).

[197] A. Locquet, C. Masoller, and C. R. Mirasso, "Synchronization regimes of optical-feedback-induced chaos in unidirectionally coupled semiconductor lasers," Phys. Rev. E **65**(5), 056205 (2002).

[198] R. Vicente, T. Perez, and C. R. Mirasso, "Open- versus closed-loop performance of synchronized chaotic external-cavity semiconductor lasers," IEEE J. Quantum Electron. **QE-38**(9), 1197–1204 (2002).

[199] I. Wedekind and U. Parlitz, "Experimental observation of synchronization and anti-synchronization of chaotic low-frequency-fluctuations in external cavity semiconductor lasers," Int. J. Bifurcation and Chaos **11**(4), 1141–1147 (2001).

[200] Y. Liu, A. Takiguchi, P. Davis, T. Aida, S. Saito, and J. M. Liu, "Experimental observation of complete chaos synchronization in semiconductor lasers," Appl. Phys. Lett. **80**(23), 4306–4308 (2002).

[201] S. Boccaletti, L. Pecora, and A. Pelaez, "Unifying framework for synchronization of coupled dynamical systems," Phys. Rev. E **63**(6), 066219 (2001).

[202] A. Uchida, R. McAllister, and R. Roy, "Consistency of nonlinear system response to complex drive signals," Phys. Rev. Lett. **93**(24), 244,102 (2004).

[203] I. V. Koryukin and P. Mandel, "Two regimes of synchronization in unidirectionally coupled semiconductor lasers," Phys. Rev. E **65**(2), 026201 (2002).

List of Publications

Contributions to Scientific Journals and Books

- M. Peil, I. Fischer, and W. Elsäßer, "Spectral Broadband Dynamics of Semiconductor Lasers with Resonant Short Cavities", Phys. Rev. A **73**, 023805, (2005).

- A. Tabaka, M. Peil, M. Sciamanna, I. Fischer, W. Elsäßer, H. Thienpont, I. Veretennicoff, K. Panajotov, "Dynamics of Vertical-Cavity Surface-Emitting Lasers in the Short External Cavity Regime: Pulse Packages and Polarization Mode Competition", Phys. Rev. A **73**, 013810, (2006).

- A. Argyris, D. Syvridis, L. Larger, V. Annovazzi-Lodi, P. Colet, I. Fischer, J. Garcia-Ojalvo, C. R. Mirasso, L. Pesquera, K. A. Shore, "Chaos-based communications at high bit rates using commercial fibre-optic links", Nature **438**, 343–346, (2005).

- H. Erzgräber, D. Lenstra, B. Krauskopf, E. Wille, M. Peil, I. Fischer, and W. Elsäßer, "Mutually Delay-Coupled Semiconductor Lasers: Mode Bifurcation Scenarios", Opt. Commun. **255**, pp. 286–296, (2005).

- M. Peil, I. Fischer, W. Elsäßer, "A Short Cavity Semiconductor Laser System: Dynamics Beyond Lang-Kobayashi", Proceedings of ENOC-2005, Fifth EUROMECH Nonlinear Dynamics Conference, D. H. van Campen, M. D. Lazurko, W. P. J. M. van den Oever, editors, ISBN 90 386 2667 3, MS 18-269, pp. 2074–2082, (2005).

- C. R. Mirasso, I. Fischer, M. Peil, and L. Larger, "Optical Chaos Communications", Proceedings of SPIE, Vol. **5825**, Optoelectronics, Photonic Devices, and Optical Networks, John G. McInerney, Gerard Farrell, David M. Denieffe, Liam Barry, Harold S. Gamble, Padraig Hughes, Robert A. Moore, editors, April 2005, pp. 139–151, (2005).

- H.-J. Wünsche, S. Bauer, J. Kreissl, O. V. Ushakov, N. Korneyev, F. Henneberger, E. Wille, H. Erzgräber, M. Peil, W. Elsäßer, I. Fischer, "Dynamics of Delay-Coupled Semiconductor Lasers: The Short Delay Regime", Phys. Rev. Lett. **94**, 163901, (2005).

- A. Tabaka, M. Peil, M. Sciamanna, I. Fischer, W. Elsäßer, H. Thienpont, I. Veretennicoff, K. Panajotov, "Experimental Studies of Pulse Packages in Short External Cavity VCSELs", Proceedings of IEEE/LEOS Benelux Chapter Annual Symposium 2004, pp. 163–166, (2005).

- M. Peil, I. Fischer, and W. Elsäßer, "A Short External Cavity Semiconductor Laser Cryptosystem", invited paper, C. R. Physique **5/6**, pp. 633–642, (2004).

- E. Wille, M. Peil, I. Fischer, W. Elsäßer, "Dynamical Scenarios of Mutually Delay-Coupled Semiconductor Lasers in the Short Coupling Regime", Proceedings of SPIE, Vol. **5452**, Semiconductor Lasers and Semiconductor Laser Dynamics, Daan Lenstra, Geert Morthier, Thomas Erneux, Markus Pessa, editors, September 2004, pp. 41–50, (2004).

- C. R. Mirasso, I. Fischer, M. Peil, L. Larger, "Optoelectronic Devices for Optical Chaos Communications", Proceedings of SPIE, Vol. **5248**, Semiconductor Optoelectronic Devices for Lightwave Communication, Joachim Piprek, editor, December 2003, pp. 24–34, (2003).

- T. Heil, J. Mulet, I. Fischer, C. R. Mirasso, M. Peil, P. Colet, and W. Elsäßer, "ON/OFF Phase Shift Keying for Chaos-Encrypted Communication using External-Cavity Semiconductor Lasers", invited paper, IEEE J. Quantum Electron. **QE-38**, pp. 1162–1170 (2002).

- M. Peil, T. Heil, I. Fischer, W. Elsäßer, "Chaos-Synchronization in Semiconductor Laser Systems: A Vectorial Coupling-Dependent Scenario", Phys. Rev. Lett. **88**, 174101, (2002).

- M. Peil, T. Heil, I. Fischer, W. Elsäßer, "Chaos-Synchronization in Semiconductor Laser Systems: An Optical Phase Dependent Scenario", in "Experimental Chaos: 6th Experimental Chaos Conference", edited by S. Boccaletti et. al., American Institute of Physics, **CP662**, pp. 433–438, (2002).

Contributions to International and National Meetings

- M. Peil, I. Fischer, W. Elsäßer, "A Short Cavity Semiconductor Laser System: Dynamics Beyond Lang-Kobayashi", ENOC-2005, Fifth EUROMECH Nonlinear Dynamics Conference, Mini-Symposium on Laser Dynamics, MS 18-269, Eindhoven, The Netherlands, 7th-12th August, (2005).

- J. von Staden, T. Gensty, M. Peil, W. Elsäßer, G. Giuliani, C. Mann, "First Investigations on the Alpha-Parameter of Quantum Cascade Lasers", COST 288 Meeting at the 2nd International Symposium on Ultrafast Photonic Technologies, St. Andrews, Scotland, UK, 1st-4th August, (2005).

- I. Fischer, E. Wille, M. Peil, W. Elsäßer, "Nonlinear Dynamics of Mutually Delay-Coupled Semiconductor Lasers", Dynamics Days 2005, M10.2 Minisymposium: Nonlinear Optical Systems (PN 202), Berlin, Germany, 25th-28th July, (2005).

- M. Peil, I. Fischer, W. Elsäßer, "Coherence Properties of a Short External Cavity Semiconductor Laser System: A Light Source With Widely Tunable Coherence Length", CLEO/Europe-EQEC Focus Meeting: New Systems, EC1-6-TUE, Munich, Germany, 12th-17th June, (2005).

- C. R. Mirasso, S. Poinsot, L. Larger, M. Peil, I. Fischer, "Communicating with Chaotic Light", invited talk, SPIE Europe Regional Meeting, Opto Ireland 2005, Optoelectronics and Photonics Devices, Conference 5825A-18, Dublin, Ireland, 4th-5th April, (2005).

- I. Fischer, E. Wille, M. Peil, W. Elsäßer, "Dynamics of Mutually Delay-Coupled Semiconductor Lasers: Scenarios in the Short Coupling Regime", PHASE 2005, International Workshop on Physics & Applications of Semiconductor Lasers, Metz, France, 29th-30th March, (2005).

- A. Tabaka, M. Peil, M. Sciamanna, I. Fischer, W. Elsäßer, H. Thienpont, I. Veretennicoff, K. Panajotov, "Experimental Studies of Pulse Packages in Short External Cavity VCSELs", PHASE 2005, International Workshop on Physics & Applications of Semiconductor Lasers, Metz, France, 29th-30th March, (2005).

- H.-J. Wünsche, S. Bauer, J. Kreissl, O. V. Ushakow, N. Korneyev, F. Henneberger, E. Wille, H. Erzgräber, M. Peil, W. Elsäßer, I. Fischer, "Synchronization and Chaos of Delay-Coupled Semiconductor Lasers: The Short Delay Regime", Optoelectronics, Physics and Simulation of Optoelectronic Devices XIII, San Jose, California, USA, Conference 5722-32, 22nd-27th January, (2005).

- A. Tabaka, M. Peil, M. Sciamanna, I. Fischer, W. Elsäßer, H. Thienpont, I. Veretennicoff, K. Panajotov, "Experimental Studies of Pulse Packages in Short External Cavity VCSELs", Proceedings of IEEE/LEOS Benelux Chapter Annual Symposium 2004, Ghent, Belgium, 2nd-3rd December, (2004).

- I. Fischer, E. Wille, M. Peil, W. Elsäßer, "Dynamics of Mutually Coupled Semiconductor Lasers: On the Role of a Coupling-Delay", invited talk, 8th Experimental Chaos Conference (ECC8), Firenze, Italy, 14th-17th June, (2004).

- E. Wille, M. Peil, I. Fischer, W. Elsäßer, "Dynamical Scenarios of Mutually Delay-Coupled Semiconductor Lasers in the Short Coupling Regime", SPIE2004, Strassbourg, France, 26th-30th October, (2004).

- E. Wille, M. Peil, I. Fischer, W. Elsäßer, "Experimente zur Dynamik zeitverzögert gekoppelter Halbleiterlaser bei kurzen Kopplungszeiten", Verhandlungen DPG (VI) **39**, DY36.1, Regensburg, Germany (2004).

- I. Fischer, M. Peil, T. Heil, W. Elsäßer, "Synchronization of Chaotic Semiconductor Lasers: Scenarios and Applications", invited talk, Frontiers in Optics, the 87th OSA Anual Meeting, Laser Science XIX, WKK2, Tucson, Arizona, 5th-9th October, (2003).

- I. Fischer, E. Wille, M. Peil, W. Elsäßer, "Dynamics of Mutually Coupled Semiconductor Lasers: Towards Small Coupling Delays", invited talk, WIAS Workshop "Dynamics of Semiconductor Lasers", Berlin, Germany, 15th-17th September, (2003).

- I. Fischer, M. Peil, T. Heil, W. Elsäßer, "Synchronization of Chaotic Semiconductor Lasers: Scenarios and Applications", invited talk, Workshop "Complex Nonlinear Processes", Berlin, Germany, 11th-13th September, (2003).

- C. R. Mirasso, I. Fischer, M. Peil, L. Larger, "Optoelectronic Devices for Optical Chaos Communications", invited talk, SPIE International Symposium, ITCom 2003, Information Technologies and Communications 2003, Semiconductor Optoelectronic Devices for Lightwave Communication, Conference 5248, Orlando, Florida, USA, 7th-11th September, (2003).

- M. Peil, T. Heil, I. Fischer, W. Elsäßer, J. M. Buldú, "Chaos-Synchronization of Semiconductor Laser Systems in an Open-Loop Configuration: The Short Cavity Regime and its Potential for Secure Communication Systems", CLEO/Europe EQEC Focus Meeting: Synchronization of Chaotic Lasers, EA4-4-FRI, Munich, Germany, 23rd-27th June, (2003).

- M. Peil, T. Heil, I. Fischer, W. Elsäßer, "Delay Synchronization of Two Unidirectionally Coupled Semiconductor Lasers with Delayed Optical Feedback: Synchronization Scenario and Potential for Communication Systems", Semiconductor and Integrated Optoelectronics, SIOE´2002 Conference, Cardiff, Wales, (2002).

- I. Fischer, M. Peil, T. Heil, W. Elsäßer, "Synchronization Phenomena of Semiconductor Lasers with Delayed Optical Feedback: The Role of the Relative Cavity Phase", invited talk, Optoelectronics, Integrated Optoelectronic Devices, San Jose, California, USA, Conference 4646, (2002).

- M. Peil, T. Heil, I. Fischer, W. Elsäßer, "Chaos-Synchronization in Semiconductor Laser Systems: A Vectorial Coupling-Dependent Scenario", seminar talk, Palma de Mallorca, Spain, (2001).

- M. Peil, T. Heil, I. Fischer, W. Elsäßer, "Synchronization of Chaotic Semiconductor Laser Systems: An Optical Phase Dependent Scenario", 6th Experimental Chaos Conference, Potsdam, Germany, (2001).

- M. Peil, T. Heil, I. Fischer, W. Elsäßer, "Chaos-Synchronisation Unidirektional Gekoppelter Halbleiterlaser", Verhandlungen DPG (VI) **36**, DY50.5, Hamburg, Germany, (2001).

Danksagungen - Acknowledgements

Abschließend möchte ich mich bei all denen recht herzlich bedanken, die mich auf meinem bisherigen wissenschaftlichen Werdegang begleitet und unterstützt haben, und somit zum Gelingen dieser Arbeit beigetragen haben. Mein Dank gilt aber auch meinen Freunden, Kommilitonen und Bekannten, mit denen ich in dieser Zeit viele schöne Stunden verbracht habe.

Ich danke Herrn *Prof. Dr. Wolfgang Elsäßer* für die fundierte fachliche Betreuung meiner Doktorarbeit, die Bereitstellung der hervorragenden Arbeitsbedingungen, sowie die Möglichkeit meine Arbeit eigenverantwortlich durchführen zu können. Weiterhin möchte ich Ihnen für die Ermöglichung der Teilnahme an zahlreichen Tagungen und Projekttreffen im In- und Ausland danken, die mir stets große Freude bereitet haben und durch die ich meinen wissenschaftlichen Horizont erweitern konnte.

Ich danke Herrn *Dr. Ingo Fischer* für die engagierte Betreuung meiner Arbeit, für wertvolle fachliche Ratschläge und Hinweise und für die Motivation zur Bearbeitung dieser interessanten Thematik. Ich danke Dir für die anregenden Diskussionen, die wir in den letzten Jahren geführt haben, aber ebenso für die kurzweiligen, kulinarischen und sportlichen Abende fernab der Physik.

Ich danke Herrn *Prof. Dr. Friedemann Kaiser* für sein Interesse an dieser Arbeit und für die Übernahme des Zweitgutachtens.

Ich danke meinen Freunden und Kommilitonen *Dr. Tobias Gensty* und *Dr. Kristian Motzek* für die facettenreichen Einblicke in die Physik jenseits des Chaos und die unterhaltsamen Dart-Abende, die ich sicherlich vermissen werde.

Ich danke den derzeitigen und ehemaligen Mitgliedern der AG Halbleiteroptik, den Herren *Dr. Christian Degen, Dr. Tilman Groth, Dr. Joachim Kaiser, Icksoon Park, Dr. Tobias Gensty, Shyam K. Mandre, Hartmut Erzgräber, Eric Wille, Klaus Becker, Richard H. Birkner, Philip Kappe, Jens von Staden, Martin Blazek, Stefan Breuer, Christian Fuchs, Markus Merkel, Saša Bakić* und *Andreas Barchanski*, für das kollegiale Arbeitsklima, die gegenseitige Unterstützung bei der Lösung messtechnischer Probleme, aber auch für die unterhaltsamen Gespräche in der Kaffee- oder mittlerweile auch in der Teepause. Mein besonderer Dank gilt Herrn *Dr. Tilmann Heil* für seine kompetente fachliche Unterstützung, die mir die Einarbeitung in die Thematik dieser Arbeit erleichtert hat.

I gratefully acknowledge the colleagues working in the field of semiconductor laser dynamics I had the pleasure to collaborate and interact with during the last years.

I would like to thank *Dr. Claudio R. Mirasso, Dr. Pere Colet, Dr. Josep Mulet,* and *Raul Vicente* for the fruitful collaboration during the last 5 years and their kind hospitality during may stay in Palma de Mallorca. I thank you for direct contribution to this thesis by providing many of the numerical results.

I am grateful to the members of the OCCULT project for exchange of knowledge and for the inspiring discussions we had at the project meetings.

I thank *Dr. Javier Martin-Buldú* and *Dr. Andrzej Tabaka* for scientific collaboration during their stays here in Darmstadt and for the fun we had doing extensive experiments till late in the night.

Weiterhin möchte ich mich bei Herrn *Dr. Joachim R. Sacher*, Frau *Dr. Sandra Stry* und Herrn *Michael Cenkier* der Firma Sacher Lasertechnik GmbH, Marburg, für die gute Zusammenarbeit und die Bereitstellung des LYNX-Systems, das als Basis zur Entwicklung des kurzen Resonator-Systems diente, bedanken.

Nicht zu letzt danke ich den wissenschaftlichen und nicht-wissenschaftlichen Mitarbeitern, die zum Gelingen dieser Arbeit beigetragen haben. Ich danke den Mitarbeitern der Mechanik- und der Elektrowerkstatt um Herrn *Karl-Heinz Vetter* und um *Herrn Günther W. Gräfe* für die fachliche Beratung, Planung und Realisierung wichtiger mechanischer und elektronischer Komponenten. Ich bedanke mich bei Herrn *Gerhard Jourdan* für die Erstellung der elektronenmikroskopischen Aufnahmen der verwendeten DFB-Halbleiterlaser, bei Frau *Barbara Hackel* für die Erstellung von Grafiken und bei Frau *Roswitha Jaschik* und Frau *Petra Gebert* für die Unterstützung bei der Erledigung verwaltungstechnischer Angelegenheiten.

Ich danke dem *Bundesministerium für Bildung und Forschung* (FK 13N8174) und der *Europäischen Kommission* (Projekt IST-2000-29683) für die Bereitstellung der finanziellen Mittel zur Förderung meiner Arbeit.

Der größte Dank gilt meinen Eltern, die mich während meiner 25-jährigen Ausbildungszeit stets moralisch und finanziell unterstützt haben. Ich danke Euch recht herzlich für diese Unterstützung und die Freiheiten, die Ihr mir während meines Studiums und der Doktorandenzeit gelassen habt. In diesem Zusammenhang möchte ich auch meiner Freundin *Kristina* danken. Ich danke Dir für die schöne Zeit mit Dir, und dafür dass Du mich in der Verwirklichung meiner Pläne so tatkräftig unterstützt. Ohne Euch wäre all dies nicht möglich gewesen.

Curriculum Vitae

address:	Michael Peil	
	Otzbergstraße 18	
	64853 Otzberg	
personal data:	date of birth:	15/1/1975
	place of birth:	Hanau am Main
	nationality:	German
school:	1985–1994	Wolfgang-Ernst-Schule (high school) (Büdingen)
	6/1994	high school graduation "Abitur"
military service:	1994–1995	with 3./FüUstgRgt 40 in Gerolstein
university:	1995–2002	studies of physics at Darmstadt University of Technology
	1998–1999	studies of physics at University of Bristol UK (ERASMUS-program)
	2000–2002	diploma thesis in the semiconductor optics group with Prof. Dr. W. Elsäßer: *Synchronization phenomena of unidirectionally coupled chaotically emitting semiconductor laser systems*
	2/2002	diploma in physics
	2002–2006	research associate and Ph.D-student in the semiconductor optics group, Institute of Applied Physics, Darmstadt University of Technology